LASERS AND ELECTRO-OPTICS RESEARCH AND TECHNOLOGY

SPECTROSCOPIC ANALYSIS OF CHEMICAL SPECIES IN CARBON PLASMAS INDUCED BY HIGH-POWER IR CO_2 LASER

LASERS AND ELECTRO-OPTICS RESEARCH AND TECHNOLOGY

Additional books in this series can be found on Nova's website under the Series tab.

LASERS AND ELECTRO-OPTICS RESEARCH AND TECHNOLOGY

SPECTROSCOPIC ANALYSIS OF CHEMICAL SPECIES IN CARBON PLASMAS INDUCED BY HIGH-POWER IR CO_2 LASER

J. J. CAMACHO
J. M. L. POYATO
L. DÍAZ
AND
M. SANTOS

Nova Science Publishers, Inc.
New York

Copyright © 2011 by Nova Science Publishers, Inc.

All rights reserved. No part of this book may be reproduced, stored in a retrieval system or transmitted in any form or by any means: electronic, electrostatic, magnetic, tape, mechanical photocopying, recording or otherwise without the written permission of the Publisher.

For permission to use material from this book please contact us:
Telephone 631-231-7269; Fax 631-231-8175
Web Site: http://www.novapublishers.com

NOTICE TO THE READER

The Publisher has taken reasonable care in the preparation of this book, but makes no expressed or implied warranty of any kind and assumes no responsibility for any errors or omissions. No liability is assumed for incidental or consequential damages in connection with or arising out of information contained in this book. The Publisher shall not be liable for any special, consequential, or exemplary damages resulting, in whole or in part, from the readers' use of, or reliance upon, this material. Any parts of this book based on government reports are so indicated and copyright is claimed for those parts to the extent applicable to compilations of such works.

Independent verification should be sought for any data, advice or recommendations contained in this book. In addition, no responsibility is assumed by the publisher for any injury and/or damage to persons or property arising from any methods, products, instructions, ideas or otherwise contained in this publication.

This publication is designed to provide accurate and authoritative information with regard to the subject matter covered herein. It is sold with the clear understanding that the Publisher is not engaged in rendering legal or any other professional services. If legal or any other expert assistance is required, the services of a competent person should be sought. FROM A DECLARATION OF PARTICIPANTS JOINTLY ADOPTED BY A COMMITTEE OF THE AMERICAN BAR ASSOCIATION AND A COMMITTEE OF PUBLISHERS.

Additional color graphics may be available in the e-book version of this book.

LIBRARY OF CONGRESS CATALOGING-IN-PUBLICATION DATA

Spectroscopic analysis of chemical species in carbon plasmas induced by
high-power IR CO2 laser / J.J. Camacho ... [et al.].
   p. cm.
 Includes index.
 ISBN 978-1-61209-575-2 (softcover)
 1. Laser plasmas. 2. Laser-induced breakdown spectroscopy. 3. Spectrum
analysis. 4. High power lasers. I. Camacho, J. J.
 QC718.5.L3S64 2011
 543'.5--dc22
                    2011003583

Published by Nova Science Publishers, Inc. † New York

Contents

Preface vii

Chapter 1 Introduction 1

Chapter 2 Fundamentals of Laser Induced Breakdown Spectroscopy (LIBS) 3

Chapter 3 Experimental 25

Chapter 4 Results and Discussion 33

Chapter 5 Conclusion 71

Acknowledgments 73

References 75

Index 79

PREFACE

This book describes some fundamentals of laser-induced breakdown spectroscopy (LIBS) and experimental results obtained from ultraviolet-visible-near infrared (UV-Vis-NIR) spectra induced by laser ablation of a graphite target, developed in our laboratory. Ablation was produced by a high-power IR CO_2 pulsed laser using several wavelengths (λ=9.621 and 10.591 µm), power density ranging from 0.22 to 6.31 GW cm^{-2} and medium-vacuum conditions (typically at 4 Pa). Spatially and time resolved analysis were carried out for the plasma plume. Wavelength-dispersed spectra of the plume reveal the emission of C, C^+, C^{2+}, C^{3+}, C^{4+}, N, H, O, N^+, O^+ and molecular features of C_2, CN, OH, CH, N_2, N_2^+ and NH. For the assignment of molecular bands a comparison with conventional emission sources was made. The characteristics of the spectral emission intensities from the different species have been investigated as functions of the ambient pressure, laser irradiance, delay time, and distance from the target. Excitation, vibrational and rotational temperatures, ionization degree and electron number density for some species were estimated. Time-gated spectroscopic studies have allowed estimation of time-of-flight (TOF) and propagation velocities for various emission species.

Chapter 1

INTRODUCTION

The interaction of a high-energy infrared (IR) pulsed laser beam with solid materials has been investigated extensively over the past several years, due to its significance in technologies such as laser sampling for chemical analysis and pulsed laser deposition of thin film [1-5]. Laser–material interactions involve non-linear complex and collective processes that are not fully understood. Among the many factors the laser irradiance is one of the most important in controlling the mechanism of the laser–material interaction. For example, at high irradiance, a laser induced plasma can be formed above the sample surface which may absorb the incident laser energy, thereby shielding the sample and decreasing the efficiency of laser energy available for mass ablation. The properties and composition of the resulting ablation plume may evolve, both as a result of collisions between particles in the plume and through plume-laser radiation interactions. The laser-target interactions will be sensitively dependent both on the nature and condition of the target material, and on the laser pulse parameters. Subsequent laser-plume interactions will also be dependent on the properties of the laser radiation, while the evolution and propagation of the plume will also be sensitive to collisions and thus to the quality of the vacuum under which the ablation is conducted and/or the presence of any background gas.

Optical emission spectroscopy (OES) is a powerful tool to get information on the laser-ablated species. For laser ablation of carbon, OES studies in different atmospheres are reported and these studies have yielded many interesting results [6-17]. The major parts of this work are already published by us in different journals [18-20].

The objectives of this work are: (1) to show some fundamentals of laser-induced breakdown spectroscopy (LIBS) and, (2) to review of our recent results on LIBS analysis of chemical species in carbon plasmas induced by high-power IR CO_2 laser, adding some new results.

OES has been used to investigate thermal and dynamical properties of a plume produced by laser ablation of a graphite target at air pressures around 4 Pa. Ablation is performed using a high-power IR CO_2 pulsed laser. The emission generated by the plasma in the spectral region 200-1100 nm is due to electronic relaxation of excited C, N, H, O, ionic fragments C^+, C^{2+}, C^{3+}, C^{4+}, and molecular features of C_2, CN, OH, NH, CH, N_2^+ and N_2. As far as we know, a spectrum so rich in atomic lines belonging to ionized species and molecular features has not been observed previously in similar experiments. Also we analyzed these spectra at different distances from the target along the plasma expansion direction. Finally we present some new results obtained from the time resolved spectroscopic analysis of the laser ablation of a graphite target. From these results temperature, electron densities and ionization degree are obtained. Also we have studied here the spectral emission intensities from different species as functions of the ambient pressure and laser irradiance.

Although OES gives only partial information about the plasma particles, this diagnostic technique helped us to draw a picture of the plasma in terms of the emitting chemical species, to evaluate their possible mechanisms of excitation and formation and to study the role of gas-phase reactions in the plasma expansion process.

Chapter 2

FUNDAMENTALS OF LASER INDUCED BREAKDOWN SPECTROSCOPY (LIBS)

Excellent textbooks and reviews about the fundamentals of laser-induced breakdown spectroscopy (LIBS) and examples of various processes are readily available today [21-24]. LIBS, also sometimes called laser-induced plasma spectroscopy (LIPS), is a technique of atomic-molecular emission spectroscopy which utilizes a highly-power laser pulse as the excitation source. LIBS can analyze any matter regardless of its physical state, being it solid, liquid or gas. Because all elements emit light when excited to sufficiently high energy, LIBS can detect different species (atomic, ionic and molecular) and limited only by the power of the laser as well as the sensitivity and wavelength range of the spectrograph/detector. Basically LIBS makes use of OES and is to this extent very similar to arc/spark emission spectroscopy. LIBS operates by focusing the laser onto a small area at the surface of the sample; when the laser is triggered it ablates a very small amount of material which instantaneously generates a plasma plume with temperatures of about 10000–30000 K. At these temperatures, the ablated material dissociates (breakdown) into excited ionic and atomic species. At the early time, the plasma emits a continuum of radiation which does not contain any known information about the species present in the plume and within a very small timeframe the plasma expands at supersonic velocities and cools. At this point the characteristic atomic/ionic and molecular emission lines of the species can be observed. The delay between the emission of the continuum and characteristic radiation is of the order of 1 µs, this is one of the reasons for temporally gating the detector. LIBS is technically similar and complementary to a number of other laser-based techniques (Raman, laser-induced fluorescence etc). In fact devices are now being manufactured which combine these techniques in a single instrument, allowing the atomic, molecular and structural characterization of a sample as well as giving a deeper insight into physical properties. A typical LIBS system consists of a pulsed laser and a spectrometer with a wide spectral range and a high sensitivity, fast response rate and time gated detector. The principal advantages of LIBS over the conventional analytical spectroscopic techniques are its simplicity and the sampling speed.

2.1. NATURE OF THE PLASMAS

Plasma is an ionized gas, a distinct fourth state of matter. The free electric charges (electrons and ions) make plasma electrically conductive (sometimes more than gold and copper), internally interactive, and strongly responsive to electromagnetic fields. Ionized gas is usually called plasma when it is electrically neutral (i.e., electron density is balanced by that of positive ions) and contains a significant number of the electrically charged particles, sufficient to affect its electrical properties and behaviour. Plasmas occur naturally but also can be effectively man-made in laboratory and in industry, which provides opportunities for numerous applications, including thermonuclear synthesis, electronics, lasers, fluorescent lamps, and many others.

Plasma offers three major features that are attractive for applications in chemical-physics: (1) temperatures and energy density of at least some plasma components can significantly exceed those in conventional technologies, (2) plasmas are able to produce very high concentrations of energetic and chemically active species (e.g., electrons, ions, atoms, molecules and radicals, excited states, and different wavelength photons), and (3) plasma systems can essentially be far from thermodynamic equilibrium, providing extremely high concentrations of chemically active species and keeping bulk temperature as low as room temperature. These plasma features permit significant intensification of traditional chemical processes, essential increase of their efficiency, and often successful stimulation of chemical reactions impossible in conventional chemistry. Plasmas are found in nature in various forms and are characterized normally by their electron density n_e and electron temperature T_e. On earth they exist in the ionosphere at height of 70-500 km (density $n_e = 10^6$ cm^{-3}, $T_e = 2300$ K). Solar wind is another natural plasma originating from the sun with $n_e = 10$ cm^{-3} and $T_e = 10^5$ K. The corona which extends around the sun has an electron density $n_e = 10^8$ cm^{-3} and its electron temperature is $T_e = 10^6$ K. Finally, white dwarfs, the final state of stellar evolution, have a n_e of 10^{30} cm^{-3}. In plasma formation, as the temperature of material is raised, its state changes from solid to liquid and then to gas. If the temperature is elevated further, an appreciable number of gas atoms are ionized and go into a high temperature gaseous state in which the charge numbers of ions and electrons are almost the same and charge neutrality is satisfied at a macroscopic scale. When the ions and electrons move collectively, these charged particles interact via Coulomb forces which are long-range forces and decay with the inverse square of the distance between charged particles. Therefore, many charged particles interact with each other by long range forces rather then through short range collision process like in a common gas. This results in different kinds of collective phenomena such as plasma instabilities and wave phenomena [25].

2.2. LOCAL THERMODYNAMIC EQUILIBRIUM (LTE). MODEL FOR THE PLASMA

Plasma description starts by trying to characterize properties of the assembly of atoms, molecules, ions and electrons rather than individual species. If thermodynamic equilibrium exits, then plasma properties can be described through the concept of temperature. Thermodynamic equilibrium is rarely complete, so physicists have settled for a useful approximation, local thermodynamic equilibrium (LTE). In LTE model it is assumed that the distribution of population densities of the electrons is determined exclusively through collisional processes and that they have sufficient rate constants so that the distribution responds instantaneously to any change in the plasma conditions. In such circumstances each process is accompanied by its inverse and these pairs of processes occur at equal rates by the principle of detailed balance. Thus, the distribution of population densities of the electrons energy levels is the same as it would be in a system in complete thermodynamic equilibrium. The population distribution is determined by the statistical mechanical law of equipartition among energy levels and does not require knowledge of atomic cross sections for its calculation. Thus, although the plasma density and temperature may vary in space and time, the distribution of population densities at any instant and point in space depends entirely on local values of density, temperature, and chemical composition of plasma. If the free electrons are distributed among the energy levels available for them, their velocities have a Maxwellian distribution

$$dn_v = n_e 4\pi \left(\frac{m}{2\pi k_B T_e}\right)^{3/2} \exp\left(-\frac{m v^2}{2k_B T_e}\right) v^2 dv, \qquad (2.1)$$

where m is the electron mass, v is the electron velocity and k_B is the Boltzmann constant. For the bound levels the distributions of population densities of neutrals and ions are given by the Boltzmann (2.2) and Saha (2.3) equations

$$\frac{N_j}{N_i} = \frac{g_j}{g_i} \exp\left(-\frac{(E_j - E_i)}{k_B T_e}\right), \qquad (2.2)$$

$$\frac{N_{z+1,k} n_e}{N_{z,k}} = \frac{g_{z+1,k}}{g_{z,k}} 2 \left(\frac{2\pi\, m\, k_B\, T_e}{h^2}\right)^{3/2} \exp\left(-\frac{Ip_{z,k}}{k_B T_e}\right), \qquad (2.3)$$

where N_i, N_j, $N_{z+1,k}$ and $N_{z,k}$ are the population densities of various levels designated by their quantum numbers j (upper), i (lower) and k (the last for the ground level) and ionic charge z and $z+1$. The term $g_{z,i}$ is the statistical weight of the designated level, E_j and E_i are the energy of the levels j and i and $Ip_{z,k}$ is the ionization potential of the ion of charge z in its ground level k. Equations (2.1)-(2.3) describe the state of the electrons in an LTE plasma. For complete LTE of the populations of all levels, including the ground state, a necessary condition is that electron collisional rates for a given transition exceed the corresponding radiative rates by about an order of magnitude [26]. This condition gives a criterion [27] for the critical electron density of the level with energy ΔE

$$n_{e,crit} \geq \frac{5}{8\sqrt{\pi}} \left(\frac{\alpha}{a_0}\right)^3 z^7 \left(\frac{\Delta E}{z^2 E_H}\right)^3 \sqrt{\left(\frac{k_B T_e}{z^2 E_H}\right)} \cong 1.6 \times 10^{12} T_e^{1/2} (\Delta E)^3, (2.4)$$

where α is fine-structure parameter, a_0 is Bohr radius, and E_H is the hydrogen ionization potential. In the final form of Eq. (2.4), $n_{e,crit}$ is given in cm^{-3}, T_e in

K and ΔE in eV. Many plasmas of particular interest do not come close to complete LTE, but can be considered to be only in partial thermodynamic equilibrium in the sense that the population of sufficiently highly excited levels are related to the next ion's ground state population by Saha-Boltzmann relations, respective to the total population in all fine-structure levels of the ground state configuration [26]. For any atom or ion with simple Rydberg level structure, various criteria were advanced for the minimum principal quantum number n_{crit} for the lowest level, often called thermal or collision limit, for which partial thermodynamic equilibrium remains valid to within 10%. One criterion with quite general validity is given by Griem [27]:

$$n_{crit} \approx \left[\frac{10 z^7}{2\sqrt{\pi} n_e}\left(\frac{\alpha}{a_0}\right)^3\right]^{2/17} \left(\frac{k_B T_e}{z^2 E_H}\right)^{1/17}. \qquad (2.5)$$

2.3. LIB Plasma

As we mentioned previously, plasma is a local assembly of atoms, molecules, ions and free electrons, overall electrically neutral, in which the charged species often act collectively. The LIB plasma is initiated by a single laser pulse. If we consider the temporal evolution of LIB plasma, at early times the ionization grade is high. As electron-ion recombination proceeds, neutral atoms and molecules form. A recombination occurs when a free electron is captured into an ionic or atomic energy level and gives up its excess kinetic energy in the form of a photon.

LIB plasmas are characterized by a variety of parameters, the most basic being the degree of ionization. A weakly ionized plasma is one in which the ratio of electrons to other species is less than 10%. At the other extreme, high ionized plasmas may have atoms stripped of many of their electrons, resulting in very high electron to atom/ion ratios. LIB plasmas typically, for low power laser intensities, fall in the category of weak ionized plasmas. At high laser power densities, LIB plasmas correspond to strong ionized plasmas.

2.3.1. Initiation Mechanism: Multiphoton Ionization (MPI) and Electron Impact Ionization (EII)

Plasma is initiated by electron generation and electron density growth. The conventional LIB plasma can be initiated in two ways: multiphoton ionization (MPI) and electron impact ionization (EII) both followed by electron cascade. MPI involves the simultaneous absorption of a number of photons n, required to equal the ionization potential $I_P(A)$ of an atom or molecule A

$$nh\nu + A \rightarrow A^+ + e + I_P(A); \; nh\nu \geq I_P(A), \tag{2.6}$$

where n is the number of photons needed to strip off an electron, which corresponds to the integer part of the quantity:

$$n = \frac{I_P + \varepsilon_{osc}}{h\nu} + 1. \tag{2.7}$$

Here ε_{osc} is the oscillation energy of a free electron in the alternating electric field. Within the classical microwave breakdown theory [28], a free electron oscillates in the alternating electric field E of the laser electromagnetic wave (with frequency ω and wavelength λ (µm)), and its oscillation energy,

$$\varepsilon_{osc}[eV] = \frac{e^2 E^2}{4m\omega^2} = \frac{e^2}{4m\pi c^3} I_W \lambda^2 = 4.67 \times 10^{-14} I_W \lambda^2, \tag{2.8}$$

remains constant. In Eq. (2.8) e is the electron charge and I_w is the laser intensity (irradiance, power density or flux density in W·cm^{-2}). The probability of MPI W_{MPI}, by absorbing simultaneously n laser photons to strip off an electron, is expressed by the classical formula [29]

$$W_{MPI}[s^{-1}] \cong \omega \; n^{3/2} \left(1.36 \frac{\varepsilon_{osc}}{I_P}\right)^n = 1.88 \times 10^{15} \lambda^{2n-1} n^{3/2} \left\{\frac{6.35 \times 10^{-14} I_W}{I_P}\right\}^n, \tag{2.9}$$

where I_P is in eV. Besides, the probability of simultaneous absorption of photons decreases with the number of photons n necessary to cause ionization.

Fundamentals of Laser Induced Breakdown Spectroscopy (LIBS)

EII process consists on the absorption of light photon by free or quasifree electrons, producing electrons with enough kinetic energy e^* to ionize one atom or molecule

$$e + nh\nu + A \rightarrow e^* + A \rightarrow 2e + A^+. \tag{2.10}$$

Two conditions must co-exist for EII to initiate: (i) an initial electron must reside in the focal volume; and (ii) the initial electron must acquire energy which exceeds the ionization energy of the material in the focus. These free or quasifree electrons can be produced by the effect of cosmic ray ionization (natural ionization), by means of MPI, or by a breakdown induced in some impurity. In air at atmospheric pressure, the natural electron density is $\sim 10^3$ cm^{-3} [30]. These electrons in the focal volume gain sufficient energy, from the laser field through inverse bremsstrahlung collision with neutrals, to ionize atoms, molecules or ions by inelastic electron-particle collision resulting in two electrons of lower energy being available to start the process again

$$e^*[\varepsilon \geq I_P(A)] + A \rightarrow A^+ + 2e; \; e^*[\varepsilon \geq I_P(A^+)] + A^+ \rightarrow A^{2+} + 2e. \tag{2.11}$$

The MPI mechanism dominates electron generation only for low exciting wavelengths. Therefore initial EII becomes a problem at a higher wavelength because neither cascade nor MPI can furnish sufficient number of electrons. At higher laser intensities, electric field of the laser is able to pull an outer shell electron out of its orbit. After the initial electron ejection the LIB plasma is commonly maintained by the absorption of optical energy and the EII. Electrons in the laser field will gain energy through electron-neutral inverse bremsstrahlung collisions and will lose energy by elastic and inelastic collisions with the neutral species through excitation of rotational and vibrational degree of freedom of molecules and excitation of electronic states. While some electrons will be lost by attachment, new electrons will be produced by ionizing collisions. At high laser intensity, few electrons will be generated with energy larger than the ionization energy. The wavelength-resolved emission spectra from the laser plasma are not expected to vary due to the plasma origin. However plasma origin may be relevant, if the enhancement is observed between UV, visible and IR excitation wavelengths. Once that LIB plasma is formed, its growth is governed by the continuity rate equation for the electron density [31]

$$\frac{dn_e}{dt} = \nu_i n_e + W_n I_W^n N - \nu_a n_e - \nu_R n_e^2 + D_e \nabla^2 n_e, \tag{2.12}$$

where ν_i is the impact ionization rate, W_n is the multiphoton ionization rate coefficient, I_w is the intensity of the laser beam, n is the number of photons required for MPI, N is the number of atoms/molecules per unit volume, ν_a is the attachment rate, ν_R is the recombination rate and D_e is the electron diffusion coefficient. The term dn_e/dt is the net rate of change in electron concentration at a point in the focal volume at a time t after the release of initiatory electrons. On the right side of the equation (2.12), the first term is the electron generation due to impact ionization. The second term on the right is MPI rate. The third, fourth and fifth terms are sink terms which represent electron attachment, recombination and diffusion, respectively. Impact ionization is defined by multiplying the number of electrons per unit volume to the impact ionization rate ν_i. The impact ionization rate refers to the rate at which electrons are generated as a result of ionizing collisions. At high laser intensity, a few new electrons can be generated and gain energy larger than their ionization energy which leads to the generation of new electrons by impact ionization, thereby leading to the cascade growth.

2.3.2. Electron Attachment, Recombination and Diffusion

Electron attachment is the rate of electron attachment ν_a multiplied by the number of electrons per unit volume. The LIB plasma typically loose electrons to the neutral species via the attachment mechanism in the form of three-body attachment or two-body dissociative attachment. Three-body attachment is: $e + AB + X \rightarrow AB^- + X$, where X appears to be a facilitator that allows the electrons to be gained by AB even through X remains unchanged throughout the process. Two-body dissociative attachment is: $e + AB \rightarrow A^- + B$. In this mechanism the electrons must exhibit a threshold electron energy that is equal to the difference between the dissociative energy of AB and the attachment energy of A, which results in the separation of A and B.

Electron recombination is the rate of electron recombination ν_R multiplied to the number of electrons per unit volume. When the electron density is high, such as during the last stage of cascade breakdown, the LIB plasma can lose electrons to ions through electron-ion recombination. Similar to the electron attachment, three-body recombination and two-body recombination occurs as:

Fundamentals of Laser Induced Breakdown Spectroscopy (LIBS)

$e + AB^+ + X \rightarrow AB + X$, $e + AB^+ \rightarrow A + B$. The electron-ion recombination rate has been studied theoretically for a three-body recombination by Gurevich and Pitaevskii [32]

$$v_R = 8.8 \times 10^{-27} \frac{n_e^2}{T_e^{3.5}} \ [s^{-1}], \tag{2.13}$$

where n_e is the electron density in cm^{-3} and T_e is the electron temperature in eV. The electron diffusion term is expressed as $D_e \nabla^2 n_e$ (Eq. (2.12)). This loss mechanism, more important for a small diameter laser beam, is the diffusion of electrons out of the focal volume. Morgan [33] referred to the combined effect of diffusion and cascade ionization as the responsible for top-hat intensity profile. By imposing an electron skin at the edge of the intensity profile, they found that the electron density grows exponentially as

$$\langle v_e \rangle = \frac{2.408 \ D_e}{a^2}, \tag{2.14}$$

where $\langle v_e \rangle$ is the average electron velocity, D_e is electron diffusion coefficient and a is the radius of the beam. The equation (2.14) is intended to be an upper boundary for diffusion losses only because laser beams typically have a radial distribution closer to the gaussian rather than top-hat distribution.

In summary, two mechanisms MPI and EII can initiate a conventional LIB plasma formation. After the LIB plasma formation the temporal growth is governed by the equation (2.12). The recombination of these two source terms (MPI and EII) and three sink terms (electron attachment, electron recombination and electron diffusion) controls the development of the conventional LIB plasma. These mechanisms that directly affect the temporal development of the LIB plasma, determine the necessary spectroscopic techniques required to spectrally resolve elemental species inside the LIB plasma.

2.4. ELEMENTS OF LIBS

In contrast to conventional spectroscopy, where one is mainly concerned with the structure of an isolated atom and molecule, the radiation from the plasma also depends on the properties of the plasma in the intermediate environment of the atomic or molecular radiator. This dependence is a consequence of the long-range Coulomb potential effects which dominate the interactions of ions and electrons with each other and with existing neutral particles. These interactions are reflected in the characteristic radiations in several ways. They can control population densities of the discrete atomic states, spectral shift and broadening by Stark effect, decrease of ionization potentials of the atomic species, cause continuum radiation emissions and emission of normally forbidden lines. Generally, the radiation emitted from a self-luminous plasma can be divided into bound-bound, bound-free, and free-free transitions.

2.4.1. Line Radiation

Line radiation from plasma occurs for electron transitions between the discrete or bound energy levels in atoms, molecules or ions. In an optically thin plasma of length l along the line of sight [34], the integrated emission intensity I_{ji} of a spectral line arising from a transition between bound levels j and i is given by

$$I_{ji} = \frac{A_{ji} h \nu_{ji}}{4\pi} \int N_j ds = h \nu_{ji} A_{ji} N_j l , \qquad (2.15)$$

where N_j is the population density of the upper level j, $h\nu_{ji}$ is the photon energy (energy difference between levels j and i), and A_{ji} is the spontaneous transition probability or Einstein A coefficient. The integration is taken over a depth of plasma viewed by the detector, and the intensity of radiation is measured in units of power per unit area per unit solid angle. Transition probabilities can be sometimes expressed via the oscillator strength f_{ji}. This is defined as the ratio of the number of classical oscillators to the number of lower state atoms required to give the same line-integrated absorption [35]. Its relationship to the Einstein coefficient is

$$f_{ji} = \frac{4\pi\varepsilon_0}{e^2} \frac{mc^3}{8\pi^2 v_{ji}} \frac{g_j}{g_i} A_{ji}. \tag{2.16}$$

The usefulness of f_{ji} is that it is dimensionless, describing just the relative strength of the transition. The detailed values of A_{ji}, g_i, and g_j can be obtained from reference compilations or from electronic databases, i.e by NIST [36].

2.4.2. Continuum Radiation

The origins of continuum radiation are both bound-free and free-free transitions. The absorption of radiation from a discrete atomic state, such that the photon has enough energy to extend above the next ionization threshold, results in a release of an electron and gives rise to the process of photoionization. The reverse process of recombination occurs when an ion and an electron recombine with emission of a photon to form an ion in the next lowest ionic state (or in the neutral atomic state). Since the upper state is continuous, the emitted or absorbed radiation in both processes is also continuous. Transitions between two free energy levels can occur in plasmas increasing the energy exchanges of charged particles. Classically, this takes place because a moving charge radiates when it is accelerated or retarded. For most cases of practical importance, these free-free transitions are classified as bremsstrahlung or cyclotron spectra. In bremsstrahlung, the acceleration of charged particle takes place via the Coulomb field of charged particles. In cyclotron radiation, the acceleration is due to the gyration of charged particles in a magnetic field. The total continuum radiation at any particular frequency $I(v)$ is the sum of the contributions from all such processes having components at the specified frequency. Thus

$$I(v)dv = \frac{1}{4\pi} \int n_e \sum_i N_i \left[\gamma(i,T_e,v) + \sum_p \alpha(i,p,T_e,v) \right] hv \ ds \ dv, \tag{2.17}$$

where $\gamma(i, T_e, v)$ is the atomic probability of a photon of frequency v being produced in the field of an atom or ion (specified by i) by an electron of mean kinetic temperature T_e making free-free transition; $\alpha(i, p, T_e, v)$ is the corresponding probability where the electron makes a free-bond transition into a level p. As before, the integration is taken over the plasma depth s.

2.4.3. Line Broadening; Determination of Electron Number Density from Stark Broadening of Spectral Lines

The shape of the spectral lines in the LIB has been studied since the first observation of the laser-induced breakdown in early 1960s. It plays an important role for the spectrochemical analysis and quantification of the plasma parameters. The observed spectral lines are always broadened, partly due to the finite resolution of the spectrometers and partly to intrinsic physical causes. In addition, the center of the spectral lines may be shifted from its nominal central wavelength. The principal physical causes of spectral line broadening are the Doppler, resonance pressure, and Stark broadening. There are several reasons for this broadening and shift. These reasons may be divided into two broad categories: broadening due to local conditions and broadening due to extended conditions. Broadening due to local conditions is due to effects which hold in a small region around the emitting element, usually small enough to assure local thermodynamic. Broadening due to extended conditions may result from changes to the spectral distribution of the radiation as it traverses its path to the observer. It also may result from the combining of radiation from a number of regions which are far from each other.

Natural Broadening

The uncertainty relates the lifetime of an excited state (due to the spontaneous radiative decay) with the uncertainty of its energy. This broadening effect results in an unshifted Lorentzian profile. The full width at half maximum (FWHM) of natural broadening for a transition with a natural lifetime of τ_{ji} is: $\Delta\lambda^N_{FWHM}=\lambda^2/\pi c \tau_{ji}$. The natural lifetime τ_{ji} is dependent on the probability of spontaneous decay: $\tau_{ji}=1/A_{ji}$. Natural broadening is usually very small compared with other causes of broadening.

Doppler Broadening

The Doppler broadening is due to the thermal motion of the emitting atoms, molecules or ions. The atoms in a gas which are emitting radiation will have a distribution of velocities. Each photon emitted will be "red" or "blue" shifted by the Doppler effect depending on the velocity of the atom relative to the observer. The higher the temperature of the gas, the wider the distribution of velocities in the gas. Since the spectral line is a combination of all of the emitted radiation, the higher the temperature of the gas, the broader will be the spectral line emitted from that gas. This broadening effect is described by a

Fundamentals of Laser Induced Breakdown Spectroscopy (LIBS)

Gaussian and there is no associated shift. For a Maxwellian velocity distribution the line shape is Gaussian, and the FWHM may be estimated as (in Å):

$$\Delta \lambda^D_{FWHM} = 7.16 \times 10^{-7} \cdot \lambda \cdot \sqrt{T/M}, \quad (2.18)$$

being λ the wavelength in Å, T the temperature of the emitters in K, and M the atomic mass in amu.

Pressure Broadening

The presence of nearby particles will affect the radiation emitted by an individual particle. There are two limiting cases by which this occurs: (i) Impact pressure broadening: The collision of other particles with the emitting particle interrupts the emission process. The duration of the collision is much shorter than the lifetime of the emission process. This effect depends on both the density and the temperature of the gas. The broadening effect is described by a Lorentzian profile and there may be an associated shift. (ii) Quasistatic pressure broadening: The presence of other particles shifts the energy levels in the emitting particle, thereby altering the frequency of the emitted radiation. The duration of the influence is much longer than the lifetime of the emission process. This effect depends on the density of the gas, but is rather insensitive to temperature. The form of the line profile is determined by the functional form of the perturbing force with respect to distance from the perturbing particle. There may also be a shift in the line center. Pressure broadening may also be classified by the nature of the perturbing force as follows: (i) *Linear Stark broadening* occurs via the linear which results from the interaction of an emitter with an electric field, which causes a shift in energy which is linear in the field strength ($\sim E$ and $\sim 1/r^2$); (ii) *Resonance broadening* occurs when the perturbing particle is of the same type as the emitting particle, which introduces the possibility of an energy exchange process ($\sim E$ and $\sim 1/r^3$); (iii) *Quadratic Stark broadening* occurs via the quadratic Stark effect which results from the interaction of an emitter with an electric field, which causes a shift in energy which is quadratic in the field strength ($\sim E$ and $\sim 1/r^4$); (iv) *Van der Waals broadening* occurs when the emitting particle is being perturbed by Van der Waals forces. For the quasistatic case, a Van der Waals profile is often useful in describing the profile. The energy shift as a function of distance is given in the wings by e.g. the Lennard-Jones potential ($\sim E$ and $\sim 1/r^6$).

Stark broadening of spectral lines in the plasma occurs when an emitting species at a distance r from an ion or electron is perturbed by the electric field. This interaction is described by the Stark effect. The linear Stark effect exists for hydrogen and for all other atoms. Stark broadening from collisions of charged species is the primary mechanism influencing the emission spectra in LIBS. Stark broadening of well-isolated lines in the plasma can be used to determine the electron number density n_e(cm^{-3}). An estimation of the Stark width (FWHM) and line shift of the Stark broadened lines is given as [26-27,34-35,37-39]:

$$\Delta\lambda^{Stark}_{FWHM} = 2W\left(\frac{n_e}{10^{16}}\right) + 3.5A\left(\frac{n_e}{10^{16}}\right)^{1/4}\left(1 - BN_D^{-1/3}\right)W\left(\frac{n_e}{10^{16}}\right), \quad (2.19)$$

$$\Delta\lambda^{Shift} = D\left(\frac{n_e}{10^{16}}\right) \pm 2A\left(\frac{n_e}{10^{16}}\right)^{1/4}\left(1 - BN_D^{-1/3}\right)W\left(\frac{n_e}{10^{16}}\right), \quad (2.20)$$

where W is the electron impact parameter or half-width, A is the ion impact parameter both in Å, B is a coefficient equal to 1.2 or 0.75 for ionic or neutral lines, respectively, D (in Å) is the electron shift parameter and N_D is the number of particles in the Debye sphere $N_D = 1.72 \times 10^9 \ T^{3/2}/n_e^{1/2}$.

The electron and the ion impact parameters are functions of temperature. The first term on the right side of Eq. (2.19) refers to the broadening due to the electron contribution, whereas the second one is the ion broadening. Since for LIB conditions Stark broadening is predominantly by electron impact, the ion correction factor can safely be neglected, and Eq. (2.19) becomes

$$\Delta\lambda^{Stark}_{FWHM} \approx 2W\left(\frac{n_e}{10^{16}}\right). \quad (2.21)$$

The coefficients W are independent of n_e and slowly varying functions of electron temperature. The minus sign in Eq. (2.20) applies to the high-temperature range of those few lines that have a negative value of D/W at low temperatures. A comprehensive list of width and shift parameters W, A and D is given by Griem [27, 39].

2.4.4. Determination of Excitation, Vibrational and Rotational Temperatures

The excitation temperature T_{exc} can be calculated according to the Boltzmann equation under the assumption of LTE. The significance of this temperature depends on the degree of equilibrium within the plasma. For plasma in LTE, any point can be described by its local values of temperature, density, and chemical composition. By considering two lines λ_{ji} and λ_{nm} of the same species, characterized by different values of the upper energy level ($E_j \neq E_n$), the relative intensity ratio can be used to calculate the plasma excitation temperature

$$T_{exc} = \frac{E_j - E_n}{k_B \ln\left[\dfrac{I_{nm} \cdot \lambda_{nm} \cdot g_j \cdot A_{ji}}{I_{ji} \cdot \lambda_{ji} \cdot g_n \cdot A_{nm}}\right]}. \tag{2.22}$$

When selecting a line pair, it is advisable to choose two lines as close as possible in wavelength and as far apart as possible in excitation energy. This is to limit the effect of varying the spectral response of the detection system. The use of several lines instead of just one pair leads to greater precision of the plasma excitation temperature estimation. In fact, though the precision of the intensity values can be improved by increasing the signal intensity, the transition probabilities A_{ji} reported in the literature exhibit significance degree of uncertainty (5-50%). The excitation temperature can be calculated from the relative intensities of a series of lines from different excitation states of the same atomic or ionic species from the slope of the Boltzmann plot $\ln[I_{ji} \cdot \lambda_{ji}/g_j \cdot A_{ji}]$ versus E_j/k_B

$$\ln\left[\frac{I_{ji} \cdot \lambda_{ji}}{g_j \cdot A_{ji}}\right] = C - \frac{E_j}{k_B \cdot T_{exc}}, \tag{2.23}$$

where I_{ji} is the emissivity (W m^{-3} sr^{-1}) of the emitted $j \rightarrow i$ spectral line, λ_{ji} is the wavelength, $g_j = 2J_j + 1$ is the statistical weight, A_{ji} is the Einstein transition probability of spontaneous emission, E_j/k_B is the normalized energy of the upper electronic level and $C = \ln(hcN_j/4\pi Q(T))$ ($Q(T)$ is the partition function). The values of the λ_{ji}, g_j, A_{ji} and E_i for selected atomic or ionic lines can be obtained from the NIST Atomic Spectral Database. A set of selected spectral lines can be chosen based on their relative strengths, accuracies and transition probabilities.

The emission spectra of the diatomic species reveal a relatively complex structure which is due to the combination of the electronic transitions from the different rotational and vibrational states [40-42]. The emission intensities of the molecular bands can be analyzed in order to calculate the molecular vibrational temperature T_{vib}. For a plasma in LTE, the intensity of an individual vibrational v'-v" band $I_{v'-v"}$ is given by

$$\ln\left(\frac{I_{v'-v"} \cdot \lambda^4_{v'-v"}}{q_{v'-v"}}\right) = A - \frac{G(v')h\,c}{k_B \cdot T_{vib}}, \qquad (2.24)$$

where A is a constant, $\lambda_{v'-v"}$ is the wavelength corresponding to the band head, $q_{v'-v"} = \left|\int_0^\infty \Psi_{v'}(R)\Psi_{v"}(R)dR\right|^2$ is the Franck-Condon factor and $G(v')hc/k_B$ is the normalized energy of the upper vibrational level. A line fit to $\ln\left(I_{v'-v"} \cdot \lambda^4_{v'-v"}/q_{v'-v"}\right)$ as a function of the upper normalized electronic-vibrational energies has a slope equal to $-1/T_{vib}$.

On the other hand, the emission intensities of the rotational lines of a vibrational band can be analyzed in order to estimate the effective rotational temperature T_{rot}. In this case it is necessary to consider the Hund's coupling case for the both electronic states implied in the transition. From the assignment of the rotational spectrum it is possible to estimate the effective rotational temperature by considering the J value for the maximum of the band $T_{rot} = (2\,B_v\,h\,c/k_B)(J_{max}+1/2)^2$, being B_v the rotational constant for upper v vibrational level and J_{max} the total angular momentum at the maximum.

Another method for estimating the vibrational and rotational temperatures is based on a simulation program of the spectra. Software developed in our laboratory [43] calculated the spectra of a diatomic molecule by summing the intensity of all rovibrational levels and convoluting the results with the instrumental line shape of the optical system. The emission intensity $I_{v',J'-v",J"}$ of a molecular line can be approximated by

$$I_{v',J'-v",J"} \approx \frac{64\pi^4 \tilde{\nu}^4_{v',J'-v",J"}}{3(2J'+1)} N_{v',J'} \overline{R}_e^2 q_{v',v"} S_{J',J"}, \qquad (2.25)$$

where $\tilde{v}_{v',J'-v'',J''}$ is the wavenumber of the transition, $2J'+1$ is the rotational degeneracy of the upper state, $N_{v',J'}$ is the population in the initial (upper) state, $\bar{R_e}$ is the average electronic transition moment, $q_{v',v''}$ is the Franck-Condon factor and $S_{J',J''}$ is the Hönl-London factor [44]. Spectrum simulations are based on comparison of experimental and calculated spectra for different rotational and vibrational population distributions which depend on temperature.

2.4.5. Ionization Degree of the Plasmas: Saha Equation

In plasma there is a continuous transition from gases with neutral atoms to a state with ionized atoms, which is determined by an ionization equation. The transition between a gas and a plasma is essentially a chemical equilibrium, which shifts from the gas to the plasma side with increasing temperature. Let us consider the first three different ionization equilibria of an atom A:

$A \leftrightarrow A^+ + e + IP(A\text{-}I)$,

$A^+ \leftrightarrow A^{2+} + e + IP(A\text{-}II)$,

$A^{2+} \leftrightarrow A^{3+} + e + IP(A\text{-}III)$.

For each ionization equilibrium, considering the atoms and ions in their ground electronic state, the LTE between ionization and recombination reactions at temperature T is described by the Saha equation (2.3)

$$\frac{n_e \cdot N_i}{N_0} = \frac{g_e \cdot g_i}{g_0} \frac{(2\pi m k_B T)^{3/2}}{h^3} e^{-E_i/k_B T}, \qquad (2.26)$$

where $n_e = N_i$ are the electron and ion densities in the different ionization equilibria in the second member of ionization equilibria. From this equation, ionization degree $n_e \cdot N_i/N_0$ can be estimated.

2.5. EFFECTS OF PHYSICAL VARIABLES IN LIBS

The variables that can influence the LIBS measurements are mainly the laser properties i.e. wavelength, energy, pulse duration, focusing spot size, shot-to-shot energy fluctuations, ambient conditions, physical properties of the sample and the detection window (delay time and gate width). How these parameters affect the precision and accuracy of LIBS are addressed below.

2.5.1. Laser Parameters

In LIBS a high-power laser is used to ablate or to breakdown a gaseous sample in the form of plasma. The primary energy related parameters influencing the laser-gas interaction are the laser peak power P_W (or radiant pulse energy per time, in W) and the laser peak intensity I_W (power density or irradiance; energy per unit area and time, W cm^{-2}) given by

$$P_W = E_W / \tau_{FWHM}, \tag{2.27}$$

$$I_W = P_W / \pi r^2, \tag{2.28}$$

where E_W (in J) is the pulse energy, τ_{FWHM} (in s) is the pulse duration at the FWHM and πr^2 is the focal spot area (cm^2). The fluence Φ_W (in J cm^{-2}) on the focused spot area, the photon flux density F_{ph} (photon cm^{-2} s^{-1}) and electric field F_E (V cm^{-1}) are given by

$$\Phi_W = E_W / \pi r^2, \tag{2.29}$$

$$F_{ph} = I_W \lambda / hc, \tag{2.30}$$

$$F_E = \sqrt{I_W / c\varepsilon_0}, \tag{2.31}$$

where λ is the laser wavelength, h is the Planck constant, c is the speed of light, and ε_0 is the electric constant. The laser peak intensity I_W, fluence, photon flux and electric field are inversely proportional to the focused spot area. For LIBS, the peak intensity I_W (and the other properties Φ_W, F_{ph}, F_E and

P_R) that can be delivered to the sample is more important than the absolute value of the laser power. For the formation of plasma, the laser fluence needs to exceed the threshold value, typically of the order of several J·cm^{-2} for a nanosecond laser pulse [45]. If the laser energy is very close to the breakdown threshold, the pulse-to-pulse fluctuations can cause the plasma condition to be irreproducible, which results in poor measurement precision. The intensities of the emission lines are proportional to the laser energy while the laser plasma is in the optical thin region. When the laser energy increases further, it produces very dense and hot plasma that can absorb laser energy. This will lead to an increase in the continuum emission and a decrease in the signal intensity. Besides, the laser pulse duration and the shot-to-shot fluctuations can also affect the signal reproducibility and hence LIBS precision.

2.5.2. Focal Properties

The laser power density at the focal volume is inversely proportional to the focused spot size. For a laser beam with a Gaussian profile, the focused beam waist w_0 is given by [46]

$$w_0 = \frac{\lambda f}{\pi w_s}, \quad (2.32)$$

where f is the focal length of the lens and w_s is the radius of the unfocused beam. The higher laser power density at the focal point can be achieved by reducing the focused beam waist using a shorter focal length lens.

On the other hand, the angular spread in laser light generated by the diffraction of plane waves passing through a circular aperture consists of a central, bright circular spot (the Airy disk) surrounded by a series of bright rings. The beam divergence angle θ, measured to edges of Airy disk, is given by $\theta = 2.44 \lambda/d$, where λ is the laser wavelength and d is the diameter of the circular aperture. It can be shown that a laser beam, with beam divergence θ_i, incident on a lens of focal length f, whose diameter is several times larger than the width of the incident beam, is focused to a diffraction-limited spot of diameter approximately equal to $f \cdot \theta_i$. If the focal region of the laser beam is assumed to be cylindrical in shape, the spot size in terms of length l, can be approximated as

$$l = (\sqrt{2} - 1)\theta_i f^2 / d. \tag{2.33}$$

2.5.3. Laser Absorption in the Plasma

In LIBS the evaporation of the material begins just after the impact of the leading edge of the laser pulse on the surface. The time required for the removal of the material is less than a nanosecond, thus once the plasma is formed, a part of the laser beam will be absorbed in the plasma by the electron-neutral or electron-ion inverse bremsstrahlung (*e-n* IB and *e-i* IB), or by photoionization (PI) of the excited states. Consequently not the full laser irradiance will be able to reach the target, which is called plasma shielding. In the case of IB absorption the free electrons gain kinetic energy from the laser beam thus promoting plume ionization and excitation through collisions with excited and ground state neutrals. The IB process is usually described by the inverse absorption length α_{IB} (cm^{-1}) [31, 47]

$$\alpha_{IB,e-n} = \sigma_{e-n}[1 - e^{-h\nu/kT}]n_e N_0, \tag{2.34}$$

$$\alpha_{IB,e-i} = n_e \sigma_{e-i} = [1 - e^{-h\nu/kT}]\frac{4e^6 \lambda^3}{3hc^4 m}\sqrt{\frac{2\pi}{3mkT}} n_e N_i, \tag{2.35}$$

where N_0 is the neutral atomic density, N_i is the ionic density, σ_{e-n} is the electron-neutral cross section of photon absorption and σ_{e-i} is the electron-ion cross section of photon absorption. The term $[1 - e^{-h\nu/kT}]$ represents the losses due to stimulated emission. The electron number density in the plasma depends on the degree of ionization, evaporation rate and the plasma expansion velocity. Moreover, the absorption coefficient shows different temperature dependence for different energy density regimes. In the case of short wavelength lasers the photoionization of the excited atoms can play significant role. In fact, the absorption coefficient of this process σ_{PI} (cm^{-1}) is given by [48]

$$\alpha_{PI} = N_n \sigma_{PI} \approx \sum_n 2.9 \times 10^{-17} \frac{(I_P)^{5/2}}{(h\nu)^3} N_n, \tag{2.36}$$

being N_n is the number density (in cm^{-3}) of the excited state n, I_p is the ionization potential in eV and $h\nu$ is the photon energy in eV. In this equation the summation is performed over the energy levels which satisfy the condition $h\nu > E_n$. Equation (2.36), although derived for hydrogen like atoms, can be applied to complex atomic systems.

Chapter 3

EXPERIMENTAL

LIBS is a plasma based method that uses instrumentation similar to that used by other spectroscopic methods (atomic emission spectroscopy, laser-induced fluorescence etc). A typical LIBS apparatus utilizes a pulsed laser that generates the powerful optical pulses used to form the plasma. Principles of laser operation in general and the operation of specific lasers are described in detail in numerous books. The discussion here will be limited to the fundamentals of the operation of the transversely excited atmospheric (TEA) carbon dioxide laser used in this work. On the other hand, it is necessary a convenient detection system. We described here the detection system used in this work.

3.1. PULSED TEA CO_2 LASER

The CO_2 laser is a near-infrared gas laser capable of very high power and with the highest efficiency of all gas lasers (≈10-20%) and for cw operation the highest output power. Although CO_2 lasers have found many applications including surgical procedure, their popular image is as powerful devices for cutting, drilling, welding or as weapons for military applications. The linear CO_2 molecule has three normal modes of vibration, labelled v_1 (symmetry stretch), v_2 (bending vibration) and v_3 (asymmetric stretch), and plotted in the upper part of Figure 1. The fundamental vibration wavenumbers are 1354, 673 and 2396 cm^{-1}, respectively. The vibrational state of the molecule is described by the number of vibrational quanta in these modes. The bending vibrational mode is twofold degenerate and can have a vibrational angular momentum along the CO_2 axis.

The number of quanta of this vibrational angular momentum is stated as an upper index to the vibrational v_2 quanta. The upper laser level (00^01) denotes the ground vibrational state for the mode v_1, the ground vibrational state for the mode v_2 which is doubly degenerate, and the first excited vibrational state for the mode v_3. The active medium is a gas discharge in a mixture of He, N_2 and CO_2. By electron impact in the discharge excited vibrational levels in the electronic ground states of N_2 and CO_2 are populated (Figure 1). The vibrational levels v = 1 in the N_2 molecule and $(v_1, v_2, v_3) = (00^01)$ in the CO_2 molecule are near-resonant and energy transfer from the N_2 molecule to the CO_2 molecule becomes very efficient. This populates the (00^01) level in CO_2 preferentially, creates inversion between the (00^01) and the (02^00) levels, and allows laser oscillations on many rotational transitions between these two vibrational states in the wavelength range 9.6-10.6 μm. The main laser transitions in CO_2 occur between the excited states of the mode $v_3(00^01)$ and the symmetric stretching mode $v_1(10^00)$ (10.6 μm) or the bending mode $v_2(02^00)$ (9.6 μm). A single line can be selected by a Littrow-grating, forming one of the resonator end mirrors.

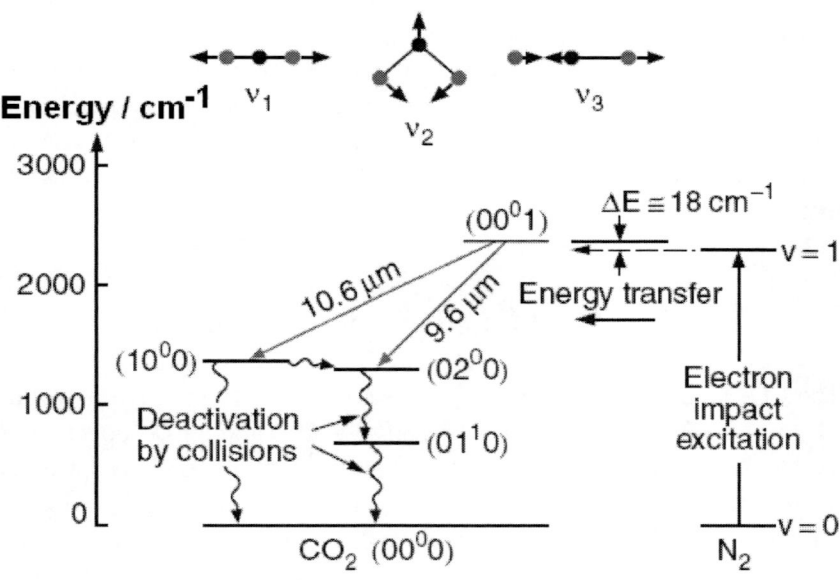

Figure 1. Level scheme and the three normal vibrational modes of the CO_2 molecule.

Helium atoms do not take part directly in the excitation of CO_2 molecules but do play an important role in heat-transfer from the gas mixture to the tube walls, as well as facilitating the depopulation of the lower vibrational levels in CO_2; contributing in this way to the ,maintenance of the population inversion. The power of CO_2 lasers depends on their configuration. The laser used in these experiments was a transversely excited atmospheric (TEA) CO_2 laser in which the electric discharge is transverse to the laser cavity's axis. The pressure in the tube is close to atmospheric pressure. The CO_2:N_2:He mixture is exchange in a continuous way, enhancing the output power of the laser. The laser can achieve a power of 50 MW. The optical materials used in lasers emitting radiation in the infrared range are obviously different than those used in the visible range. For example, materials such as germanium (Ge) or gallium arsenide (GaAs) are completely opaque in the visible range, while being transparent in the infrared range. Some materials, such as zinc selenide (ZnSe), are transparent in both spectral ranges. Typical materials transparent in the IR range are: NaCl or CsI. Metal mirrors (copper, molybdenum, gold) are used in the IR range, owing to their small absorption (and large reflectivity) as well as their large heat capacity which enables removal of heat from the active medium.

3.2. SPECTROGRAPHS AND DETECTORS

A detection system consists in a wavelength dispersing element and an electronic device as detector. Wavelength dispersing elements used for LIBS generally have to fulfilled two opposite requests: (i) need for high resolving power $\lambda/\Delta\lambda$ because of highly pronounced spectral interferences in atomic-molecular emission spectroscopy; (ii) For an analysis with a single laser pulse the wide spectral region should be covered. Today, usually, four different spectrograph mountings are mainly in use for OES: Czerny-Turner, Echelle, Rowland and Paschen-Runge. Here, only the first system used by us will be described. A Czerny-Turner mount uses a plane grating. The incident radiation passes through the entrance slit and strikes a parabolic collimating mirror. This mirror produces a collimated light beam reflected onto the grating, which spatially disperses the spectral components of the incident radiation. Collimated rays of diffracted radiation strikes a second parabolic focusing mirror. The dispersed radiation is focused in the focal plane producing the entrance slit images in that plane. Because the parallel rays of a given wavelength are incident on the focusing element at a specific angle, each wavelength is focused to a slit image at a different center position on the focal plane. In this focal plane the detector is placed (see Figure 2).

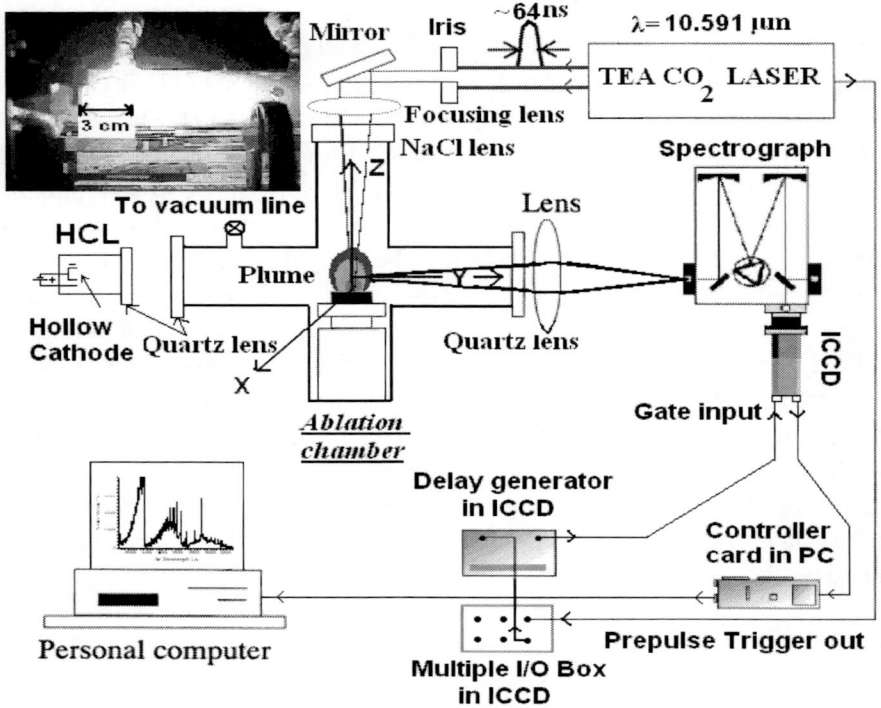

Figure 2. Schematic diagram of the experimental set-up for pulsed laser ablation diagnostics.

For the detection of plasma emissions in LIBS experiments, photomultipliers (PMT), charge-coupled and intensified charge coupled devices (CCD and ICCD) and optical multichannel analyzers (OMA) can be utilized depending on the type of spectrometer. A PMT consists of the photo-emissive cathode, several dynodes, and an anode. When the radiation hits the chatted electrons are generated. A high voltage across the dynodes leads to further multiplication of these photoelectrons. A PMT is relatively inexpensive and very sensitive detector, which allows recording of time-integrated plasma emission signals. It requires, however, a scanning mode monochromator to obtain full spectral information. Therefore, the PMT is not convenient for multielement analysis with a single laser pulse or when fast scan results are required. The charge coupled devices (CCDs) are made up of a two-dimensional array of semiconductor capacitors (pixels) that have been formed on a single silicon chip. Commercially available CCD chips usually consisted of 512×512 or 1024×1024 pixels. Performance characteristics of these instruments with respect to sensitivity, dynamic range, and signal-to-noise ratio appear to approach those of PMT's in addition of having the multichannel advantage. A CCD system can be equipped with an intensifier (microchannel plate) which allows high photonics gain and possibility of electronic gating which is of essential importance for time resolved LIBS measurements.

3.3. SCHEMATIC DIAGRAM FOR LIBS

The experimental configuration used to study carbon by LIBS is shown in Figure 2. The laser-induced plasma was generated using a TEA CO_2 laser (Lumonics model K-103) operating on an 8:8:84 mixture of CO_2:N_2:He, respectively. The laser is equipped with frontal Ge multimode optics (35 % reflectivity) and a rear diffraction grating with 135 lines mm^{-1} blazed at 10.6 μm. The CO_2 laser irradiation of the target was carried out using the 9P(28) line at λ=9.621 μm (ΔE=0.1289 eV) and the 10P(20) line at λ=10.591 μm (ΔE=0.1171 eV). The temporal shape of the TEA-CO_2 laser pulse, monitored with a photon drag detector (Rofin Sinar 7415), consisted in both cases in a prominent spike of a full width at half maximum (FWHM) of around 64 ns carrying ~90% of the laser energy, followed by a long lasting tail of lower energy and about 3 μs duration. The laser pulse repetition rate was usually 1 Hz. The divergence of the emitted laser beam is 3 mrad. The pulsed CO_2 laser beam was focused with a NaCl lens of 24 cm focal length onto the target. The primary laser beam was angulary defined and attenuated by a diaphragm of 17.5 mm diameter before entering to the cell. The CO_2 laser average energy was measured in front of the lens with a Lumonics 20D pyroelectric detectors through a Tektronix TDS 540 digital oscilloscope. Energy losses were estimated by making pulse energy measurements with and without the NaCl window in place. The focused radius of the laser beam (0.05 cm) was measured at the target position with a pyroelectric array Delta Development Mark IV. The laser intensity (power density or irradiance) on the focal volume was varied using several calibrated CaF_2 attenuators and range from 0.29 to 6.31 GW×cm^{-2}. The high purity graphite target (~99.99 %) was placed in a low-vacuum cell equipped with a NaCl window for the laser beam and two quartz windows for optical access. The graphite target is initially at ambient temperature (298 K) and it is not water-cooled. The cell was evacuated with the aid of a rotary pump, to a base pressure of 4 Pa that was measured by a mechanical gauge. Optical emission from the plume was imaged by a collecting optical system onto the entrance slit of different spectrometers. The light emitted from the laser-induced plasma was optically imaged 1:1, at right angles to the normal of the target surface, by a quartz lens (focal length 4 cm, f-number = $f/2.3$) onto the entrance slit of the spectrometer. The distance between plasma axis and entrance slit was 16 cm. The lens causes a bit chromatic aberration for OES measurements, although the geometric efficiency is barely affected. Two spectrometers were used: 1/8 m Oriel spectrometer (10 and 25 μm slits) with two different gratings (1200 and 2400 grooves mm^{-1}) in the spectral region 2000-11000 Å at a resolution of ~1.3 Å in first-order (1200 grooves mm^{-1} grating), and an ISA Jobin Yvon Spex (Model HR320) 0.32 m equipped with a plane holographic grating (2400 grooves mm^{-1}) in the spectral region 2000-7500 Å at a resolution of ~0.12 Å in first-order.

Two detectors were attached to the exit focal plane of the spectrographs and used to detect the optical emissions from the laser-induced plasma: an Andor DU420-OE (open electrode) CCD camera (1024$\square$256 matrix of 26$\square$26 μm^2 individual pixels) with thermoelectric cooling working at –30 °C; A 1024$\square$1024 matrix of 13$\square$13 μm^2 individual pixels ICCD (Andor iStar DH-734), with thermoelectric cooling working at –20 °C. The low noise level of the CCD allows long integration times and therefore the detection of very low emission intensities. The intensity response of the detection system was calibrated with a standard (Osram No.6438, 6.6-A, 200-W) halogen lamp and Hg/Ar pencil lamp. Several (Cu/Ne, Fe/Ne and Cr/Ar) hollow cathode lamps (HCL) were used for the spectral wavelength calibration of the spectrometers.

3.4. TIMING CONSIDERATIONS

For synchronization, the CO_2 laser was operated at the internal trigger mode and the ICCD detector in external and gate modes. The external trigger signal generated by the laser is fed directly into the back of the ICCD detector head. The total insertion delay (or propagation delay) is the total length of time taken for the external trigger pulse to travel through the digital delay generator and gater so that the ICCD will switch on. This insertion delay time is 45 ± 2 ns. The time jitter between the laser and the fast ICCD detector gate was about ± 2 ns. The delay time t_d is the time interval between the arrival of the laser pulse on the target and the activation of the ICCD detector. The gate width delay time t_w is the time interval during which the plasma emission is monitored by the ICCD.

Both parameters were adjusted by the digital delay generator of the ICCD detector. The resolution of the gate pulse delay time and the gate pulse width time are 25 ps. The CO_2 laser pulse picked up with the photon drag detector triggers a Stanford DG 535 pulse generator which is used as external trigger in the ICCD camera. The laser pulse and the gate monitor output were displayed in a Tektronix TDS 540 digital oscilloscope, allowing to control t_d eliminating the insertion time of the camera.

Chapter 4

RESULTS AND DISCUSSION

When a high-power laser pulse is focused on a solid surface the target becomes ablated. If the laser irradiance in the focal volume surpasses the breakdown threshold of the system formed by the vaporized atoms and residual gas, a breakdown, characterized by a brilliant flash of light accompanied by a distinctive cracking noise, is produced. At the top of figure 2 we show an image of laser-induced breakdown (LIB) plasma in graphite induced by a single CO_2 laser pulse. The plume size is around 14 cm. The laser was focused on a point at the centre of the target. The observations of the LIB geometry during the experiments indicate that the actual plasma region is not entirely spherical, but lightly elongated in the direction of the laser beam propagation. The CO_2 laser pulse remains in the focal volume after the plasma formation for some significant fraction of its duration and the plasma formed can be heated to very high temperatures and pressures by inverse bremsstrahlung absorption. Since plasmas absorb radiation much more strongly than ordinary mater, plasmas can block transmission of incoming laser light to a significant degree; a phenomenon known as plasma shielding [49]. The high temperatures and pressures produced by plasma absorption can lead to thermal expansion of the plasma at high velocities, producing an audible acoustic signature, shock waves, and cavitation effects. The plasma also tends to expand back along the beam path toward the laser, a phenomenon known as moving breakdown. The shock wave heats up the surrounding gas which is instantaneously transformed in strongly ionized plasma.

For the present experiments the measured focused-spot area was 7.85×10^{-3} cm^2. This value is higher than the calculated area (2.2×10^{-4} cm^2) obtained from the beam waist (Eq. 2.32). This fact is due to the non-gaussian profile of the

CO_2 laser beam. Moreover the CO_2 laser beam passes through a circular aperture of diameter 17.5 mm. For this diaphragm the calculated divergence angle for the laser beams at 9.621 and 10.591 μm are 1.3 and 1.5 mrad, respectively. Thus, considering the total beam divergence (~4.4 mrad), the calculated diameter of the focused TEA-CO_2 laser (beam waist) is 1.06 mm, which is very similar to the measured value (~1 mm). If the focal region of the laser beam is assumed to be cylindrical in shape, the spot size in terms of length l (Eq. 2.33) of the focused TEA-CO_2 laser is 6.0 mm, which is similar to the measured value (~7 mm).

Table 1. Laser parameters for some LIBS experiments

Energy E_W (mJ)	Power P_W (MW)	Intensity I_W (GW cm^{-2})	Fluence Φ_W (J cm^{-2})	Photon Flux, F_{ph} (photon cm^{-2} s^{-1})	Electric Field F_E (MV cm^{-1})
3161	49.5	6.31	403	3.07×10^{29}	1.54
2685	42.1	5.36	342	2.60×10^{29}	1.42
2256	35.4	4.50	287	2.18×10^{29}	1.30
1732	27.1	3.46	220	1.67×10^{29}	1.14
1209	19.0	2.41	154	1.17×10^{29}	0.985
503	7.88	1.00	64.0	4.86×10^{28}	0.615
324	5.08	0.648	41.3	3.14×10^{28}	0.494
273	4.27	0.544	34.7	2.64×10^{28}	0.453
242	3.79	0.483	30.8	2.34×10^{28}	0.427
203	3.19	0.406	25.9	1.97×10^{28}	0.391
171	2.68	0.341	21.8	1.65×10^{28}	0.358
149	2.33	0.297	19.0	1.44×10^{28}	0.335
131	2.08	0.262	16.7	1.26×10^{28}	0.314
110	1.73	0.220	14.0	1.07×10^{28}	0.288

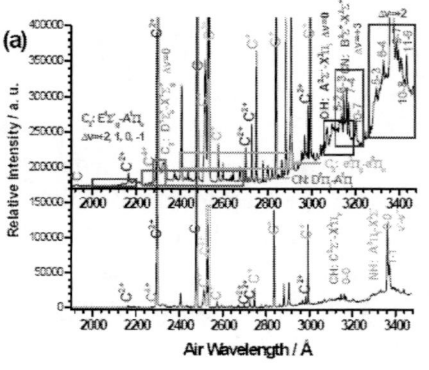
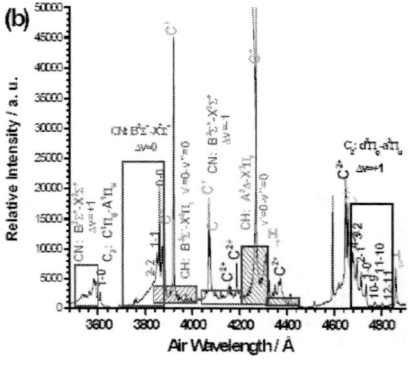

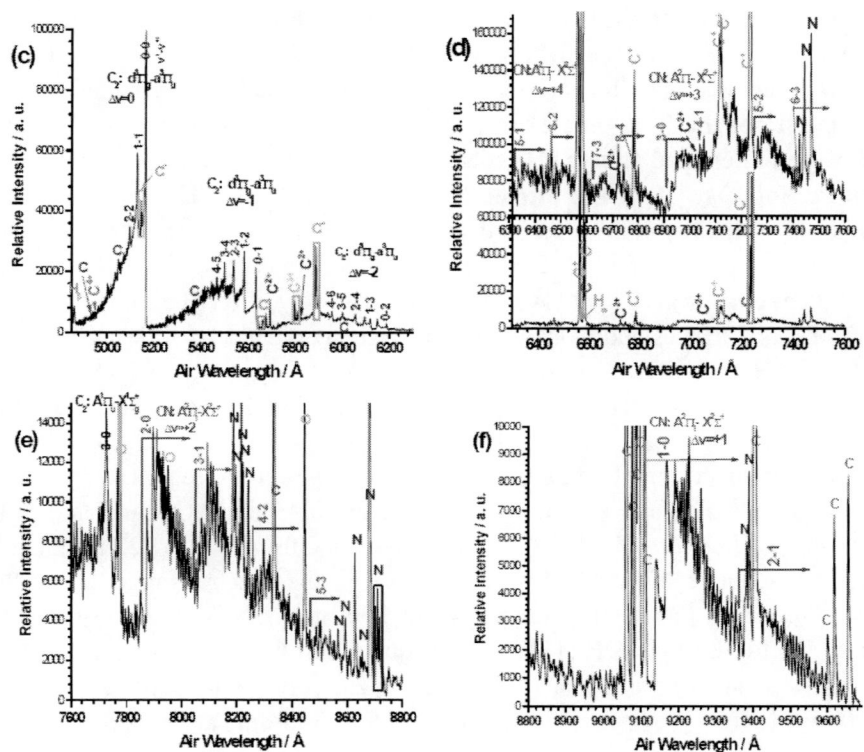

Figure 3. (a)-(f). Low-resolution PLA of carbon emission spectrum observed in the 1920-9680 Å region at an air pressure of 4 Pa, excited by the 10P(20) line at 944.20 cm^{-1} of the CO_2 laser, and assignment of the atomic lines of C, C^+, C^{2+}, C^{3+}, C^{4+}, N, O and molecular bands of $C_2(E^1\Sigma_g^+-A^1\Pi_u$; Freymark system), $C_2(D^1\Sigma_u^+ - X^1\Sigma_g^+$; Mulliken system), $C_2(e^3\Pi_g - a^3\Pi_u$; Fox-Herzberg system), $CN(D^2\Pi_i-A^2\Pi_i)$, $OH(A^2\Sigma^+-X^2\Pi)$, $CH(C^2\Sigma^+-X^2\Pi)$, $CN(B^2\Sigma^+-X^2\Sigma^+$; Violet system), $NH(A^3\Pi-X^3\Sigma^-)$, $C_2(C^1\Pi_g- A^1\Pi_u$; Deslandres-d'Azambuja system), $C_2(d^3\Pi_g-a^3\Pi_u$; Swan band system), $CH(B^2\Sigma^--X^2\Pi)$, $CH(A^2\Delta-X^2\Pi)$, $C_2(A^1\Pi_u- X^1\Sigma_g^+$; Phillips system), and $CN(A^2\Pi-X^2\Sigma^+$; red system).

4.1. IDENTIFICATION OF THE CHEMICAL SPECIES IN THE PULSE LASER ABLATION PLASMA PLUME

For the different pulse laser energies measured in this work, the calculated laser peak power (Eq. 2.27), intensity (Eq. 2.28), fluence (Eq. 2.29), photon flux (Eq. 2.30), and electric field (Eq. 2.31) are tabulated in Table 1. In the scanned spectral region, from UV to NIR, OES reproduce particular emission of carbon plasmas in a low-vacuum air atmosphere (P_{air}=4 Pa). Typical time-integrated and spectral-resolved low-resolution OES from LIB of graphite is shown in Figure 3(a)-(f). In the recording of the spectra of Figure 3(c-f) a cutoff filter was used in order to suppress high diffraction orders. In general, the spectra of the PLA plume are dominated by emission of strong electronic relaxation of excited atomic C, ionic fragments C^+, C^{2+} and C^{3+}, and molecular features of $C_2(d^3\Pi_g - a^3\Pi_u$; triplet Swan band system). The medium-weak emission is mainly due to excited atomic N, H, O, ionic fragment C^{4+} and molecular features of $C_2(E^1\Sigma_g^+ - A^1\Pi_u$; Freymark system), $C_2(D^1\Sigma_u^+ - X^1\Sigma_g^+$; Mulliken system), $CN(D^2\Pi - A^2\Pi)$, $C_2(e^3\Pi_g - a^3\Pi_u$; Fox-Herzberg system), $C_2(C^1\Pi_g - A^1\Pi_u$; Deslandres-d'Azambuja system), $OH(A^2\Sigma^+ - X^2\Pi)$, $CH(C^2\Sigma^+ - X^2\Pi)$, $NH(A^3\Pi - X^3\Sigma^-)$, $CN(B^2\Sigma^+ - X^2\Sigma^+$; violet system), $CH(B^2\Sigma^- - X^2\Pi)$, $CH(A^2\Delta - X^2\Pi)$, $C_2(A^1\Pi_u - X^1\Sigma_g^+$; Phillips system) and $CN(A^2\Pi - X^2\Sigma^+$; Red system).

In the spectrum of Figure 3(a) in the 1920-3480 Å region, very strong atomic C, C^+, C^{2+} and C^{3+} lines dominate, but also weak C^{4+} and molecular bands of $C_2(E-A$; $\Delta v=v'-v''=+2, +1, 0, -1$ sequence from 200 to 222 nm), $C_2(D-X$; $\Delta v=0$ sequence near 231.4 nm), $CN(D-A$; in the spectral range 223 to 260 nm), $C_2(e-a$; in the spectral range 240 to 290 nm), $CN(B-X$; $\Delta v=3$ sequence from 306 to 326 nm), $OH(A-X$; $\Delta v=0$ sequence from 306 to 318 nm), $CH(C-X$; $\Delta v=0$ sequence from 314 to 317 nm), $NH(A-X$; $\Delta v=0$ sequence near 336 nm) and $CN(B-X$; $\Delta v=2$ sequence from 326 to 348 nm) are observed. In this spectrum the predominant emitting species are the C^{2+} $2p^2\ ^1D_2 \rightarrow 2s2p\ ^1P_1^0$ atomic line at 2296.87 Å, C $2p(^2P_0)3s\ ^1P_1^0 \rightarrow 2p^2\ ^1S_0$ atomic line at 2478.56 Å, two lines of C^{3+} at 2524.41 and 2529.98 Å, several lines of C^+ at 2836.71 and 2992.62 Å and the $v'=0-v''=0$ band of $NH(A-X)$ at 3360 Å. In the spectrum of Figure 3(b), the predominant emitting species are C^+ (doublet $2s^24s\ ^2S_{1/2} \rightarrow 2s^23p\ ^2P_{1/2,3/2}^0$ at 3918.98 and 3920.69 Å, respectively, and multiplet $2s^24f\ ^2F_{J'}^0 \rightarrow 2s^23d\ ^2D_{J''}$ around 4267 Å), and the molecular bands of $CN(B-X$; $\Delta v=0$ sequence). Many medium intensity atomic lines of C^+, C^{2+} and C^{3+}, weak hydrogen lines of the Balmer series (H_β, H_γ etc), and several molecular bands of CN, C_2, and CH are also present. In the spectrum of Figure 3(c), the predominant emitting species are C^+ and C_2 (molecular bands: d-a; $\Delta v=0, -1$, and -2 sequences from 480 to 630 nm). Many weak lines of C, C^+, C^{2+} and C^{3+} are also present. In the spectrum of Figure 3(d), the most intense lines are the doublet structure of C^+ $2s^23p\ ^2P_{3/2,1/2}^0 \rightarrow 2s^23s\ ^2S_{1/2}$ at 6578.05 and 6582.88 Å, respectively, C $2s^22p(^2P_0)4d\ ^1P_1^0 \rightarrow 2s^22p(^2P_0)3p\ ^1P_1$ atomic line at 6587.61 Å, C^+ $2s^2 3d\ ^2D_{3/2} \to 2s^2 3p\ ^2P^0_{1/2}$ at 7231.32 Å and C^+ $2s^2 3d\ ^2D_{5/2} \to 2s^2 3p\ ^2P^0_{3/2}$ at 7236.42 Å. Also many weak lines of C, C^+, C^{2+}, H_α, N, and several bands v'-v'' (5-1, 6-2, 7-3, 8-4, 3-0, 4-1, 5-2 and 6-3) corresponding to CN(A–X) are also present. The spectrum of Figure 3(e), shows the emission of many atomic lines of C, O, and N, the 3-0 band of C_2(A-X) and several bands (2-0, 3-1, 4-2, and 5-3) of CN(A-X). Finally, in the spectrum of Figure 3(f), the emission of many atomic lines of C and N and mainly the 1-0 and 2-1 bands of CN(A–X) can be appreciated.

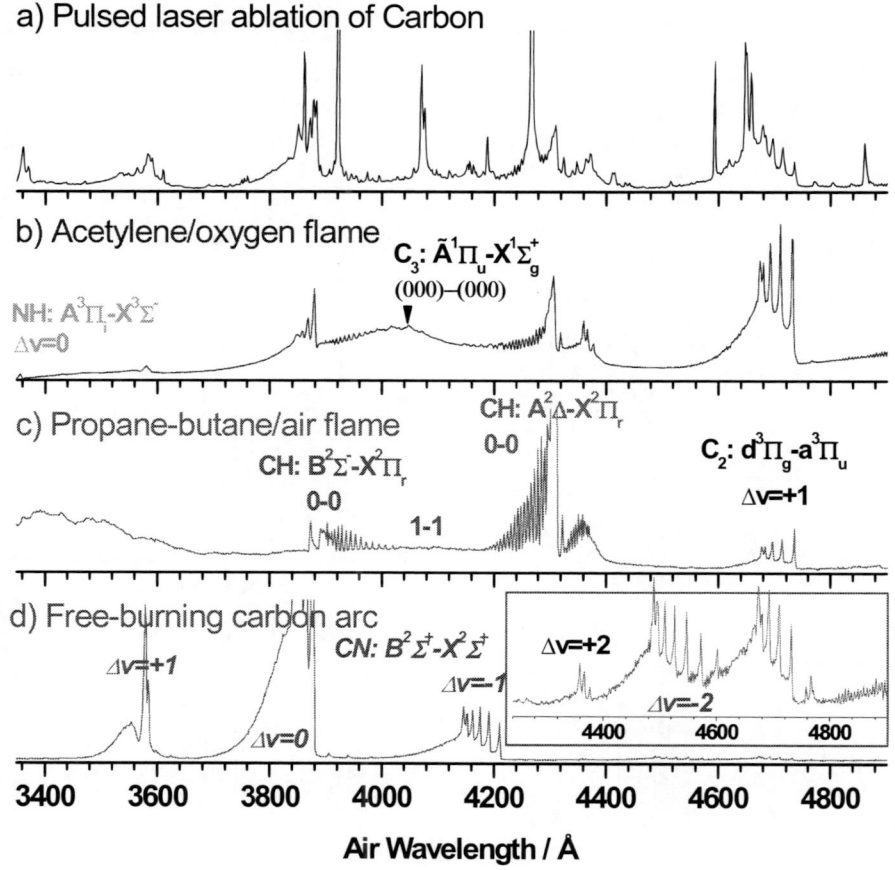

Figure 4. Low-resolution emission spectra from: a) PLA of carbon at an air pressure of 4 Pa, excited by the 9P(28) line at 1039.36 cm^{-1} of the CO_2 laser; b) Acetylene/oxygen flame; c) Propane-butane/air flame; d) Free-burning carbon arc.

For the assignment of the atomic lines of C, C^+, C^{2+}, C^{3+}, C^{4+}, H, N and O we used the information tabulated in NIST Atomic Spectral Database [36]. The observed emission molecular bands are identified using the spectroscopic information available in Refs. [50]. Moreover, these molecular bands were compared with the spectra obtained in our laboratory by conventional sources (free-burning carbon arc, propane-butane/air flame and acetylene/oxygen flame). As example figure 4 shows several time-integrated OES at low-resolution from: (a) PLA of carbon (air pressure of 4 Pa and CO_2 laser power density I_W=1.00 GW cm^{-2}); (b) Acetylene/oxygen flame; (c) Propane-butane/air flame; (d) Free-burning carbon arc. In the Acetylene/oxygen flame around 405 nm, several bands of the C_3($\tilde{A}^1\Pi_u - \tilde{X}^1\Sigma_g^+$) comet head group are observed which were not detected in the PLA of carbon. As shown in Figure 4, ionic carbon lines C^+, C^{2+}, C^{3+} and C^{4+} cannot be observed in flames [Figure 4(b,c)] or carbon electric arcs [Figure 4(d)].

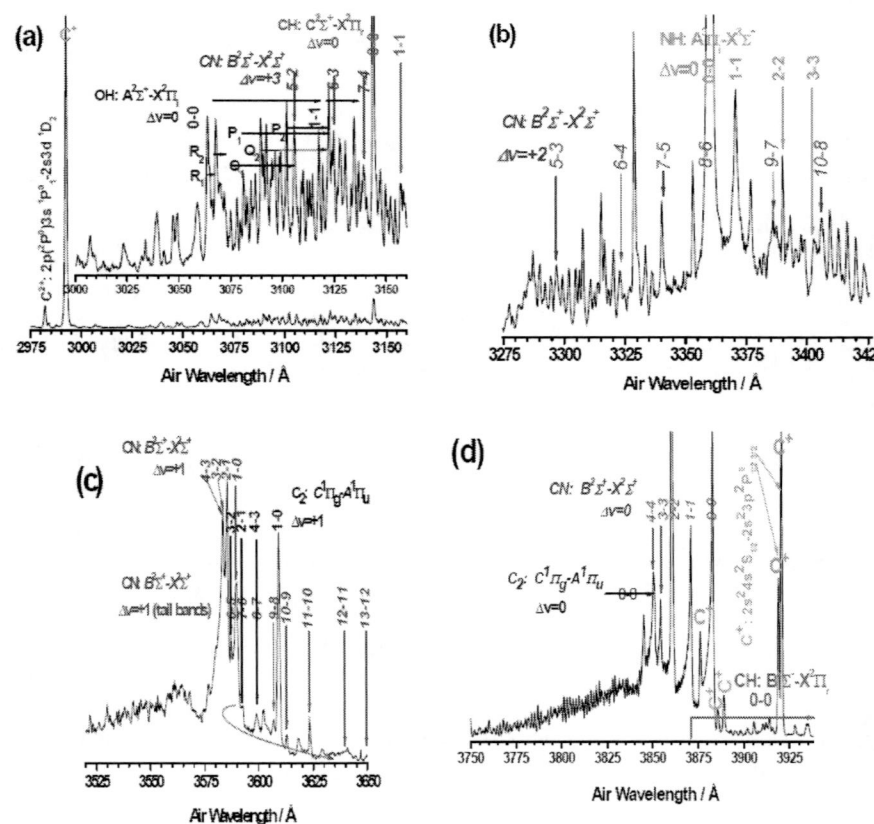

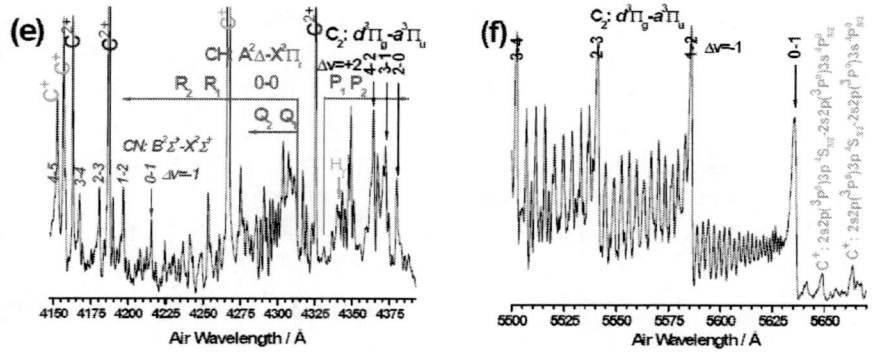

Figure 5. (a)-(f). Measured high-resolution PLA of carbon emission spectra observed in different regions at an air pressure of 4 Pa, excited by the 9P(28) line of the CO_2 laser with a laser intensity of 5.36 GW cm^{-2}, and assignment of some atomic lines and molecular band heads.

In order to get more insight into PLA of graphite and to obtain an unambiguous assignment of the emission lines and molecular bands, we have scanned the corresponding wavelength regions with higher resolution (~0.12 Å in first-order). The spectra have been obtained with twenty-four successive exposures on the CCD camera in the spectral region 200-750 nm by a ISA Jobin Yvon Spex 0.32 m spectrometer. As examples, Figure 5(a-f) shows several spectra recorded in the PLA of carbon experiment at high-resolution. These spectra were recorded in the following experimental conditions: air pressure 4 Pa, CO_2 laser excitation line 9P(28) at 9.621 μm and laser intensity 5.36 GW cm^{-2}. The relative intensities of the observed emission lines reasonably agree with tabulated values in NIST Atomic Spectral Database [36]. In Figure 5(a-f) we have indicated with italic the position of the band heads v'-v" of violet system of CN while in regular typeface the bands of the other molecular systems. In Figure 5(a-f), a rather complex structure is observed, in consequence of the overlapping between rotational lines of different molecular band systems. Figure 5(a) displays the overlapping between CH($C-X$; Δv=0 sequence), CN($B-X$; Δv=3 sequence), and OH($A-X$; Δv=0 sequence). The relative position of the main branches for the OH($A-X$) 0-0 band is indicated. In Figure 5(b), the high intensity of the 0-0 band for NH($A-X$) is observed. This fact is in agreement with the high Franck-Condon factor (q_{00}=0.9998) for this transition. In Figure 5(c) a partial overlapping among CN($B-X$; Δv=1) and C_2($C-A$; Δv=1) is observed. This spectrum clearly shows the reversal of the bands from v"=5, which is due to the overlap between high vibrational quantum number bands with low vibrational quantum number bands. So, the first vibrational bands (1-0, 2-1, 3-2, 4-3 and 5-4) are shaded to the violet and after reversal (6-5, 7-6, ...) are shaded to the red. Figure 5(d) shows a portion of the rotational lines for the CH(B-X) 0-0 band with several single ionized carbon lines. A coincidence in the position among the CN(B-X) 4-4 and C$_2$(C-A) 0-0 band heads is observed. Very weak emission attributable to the N$_2^+$($B^2\Sigma_u^+$–$X^2\Sigma_g^+$) system (the most prominent v'=0-v''=0 transition appears at ~ 391 nm) is also identifiable. In the spectrum of Figure 5(e) the CN(B-X) Δv=-1 sequence, CH(A-X) 0-0 band, and C$_2$(d-a) Δv=2 sequence were identified. Also, several C$^+$, C^{2+}, and atomic hydrogen lines are observed. Finally, Figure 5(f) displays the rotational structure of C$_2$(d-a) 0-1, 1-2, and 2-3 bands. The spectral features clearly show the complexity of the relaxation process and bring out the possibility of cascading processes.

4.2. Plasma Excitation, Vibrational and Rotational Temperature Measurements

The excitation temperature T_{exc} was calculated from the relative intensities of some C$^+$ atomic lines (250-470 nm spectral region) according to the Boltzmann equation (2.23). The estimated excitation temperature was T_{exc}= 23000 ± 1900 K (Figure 6). The relevant spectroscopic parameters for the C$^+$ transitions have been listed in Table 2.

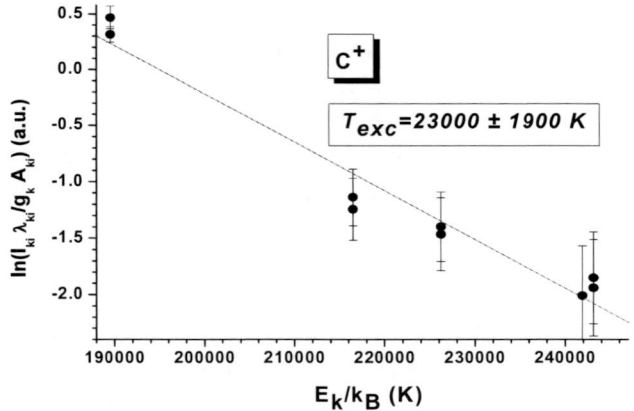

Figure. 6. Linear Boltzmann plot for several C$^+$ transition lines used to calculate plasma temperature, T_{exc}. Plot also shows linear fit to the data with a regression coefficient of R^2~0.98.

Table 2. List of C$^+$ transition lines and their spectral database
(NIST Atomic Spectra Database, 2006) used for
plasma temperature calculation

Transition array	Air λ (Å)	g_i	g_j	A_{ji} (s^{-1})	E_i (cm^{-1})	E_j (cm^{-1})	Rel. Int. (Arb. Uni.)
2s2p^2 ^{2}P$_{1/2}$–2p^3 ^{2}D$^0_{3/2}$	2509.12	2	4	4.53×10^7	110624.17	150466.69	20795
2s2p^2 ^{2}P$_{3/2}$–2p^3 ^{2}D$^0_{5/2}$	2512.06	4	6	5.42×10^7	110665.56	150461.58	41490
2s^{2}3p ^{2}P$^0_{1/2}$–2s^2 4d ^{2}D$_{3/2}$	2746.49	2	4	4.36×10^7	131724.37	168123.74	8500
2s2p^2 ^{2}S$_{1/2}$–2s^{2}3p ^{2}P$^0_{3/2}$	2836.71	2	4	3.98×10^7	96493.74	131735.52	76920
2s2p^2 ^{2}S$_{1/2}$–2s^{2}3p ^{2}P$^0_{1/2}$	2837.60	2	2	3.97×10^7	96493.74	131724.37	44700
2s^{2}3p ^{2}P$^0_{1/2}$–2s^2 4s ^{2}S$_{1/2}$	3918.98	2	2	6.36×10^7	131724.37	157234.07	7500
2s^{2}3p ^{2}P$^0_{3/2}$–2s^2 4s ^{2}S$_{1/2}$	3920.69	4	2	1.27×10^8	131735.52	157234.07	16000
2s^{2}3d ^{2}D$_{3/2}$–2s^2 4f ^{2}F$^0_{5/2}$	4267.00	4	6	2.23×10^8	145549.27	168978.34	45000
2s^{2}3d ^{2}D$_{5/2}$–2s^2 4f ^{2}F$^0_{7/2}$	4267.26	6	8	2.38×10^8	145550.70	168978.34	70000

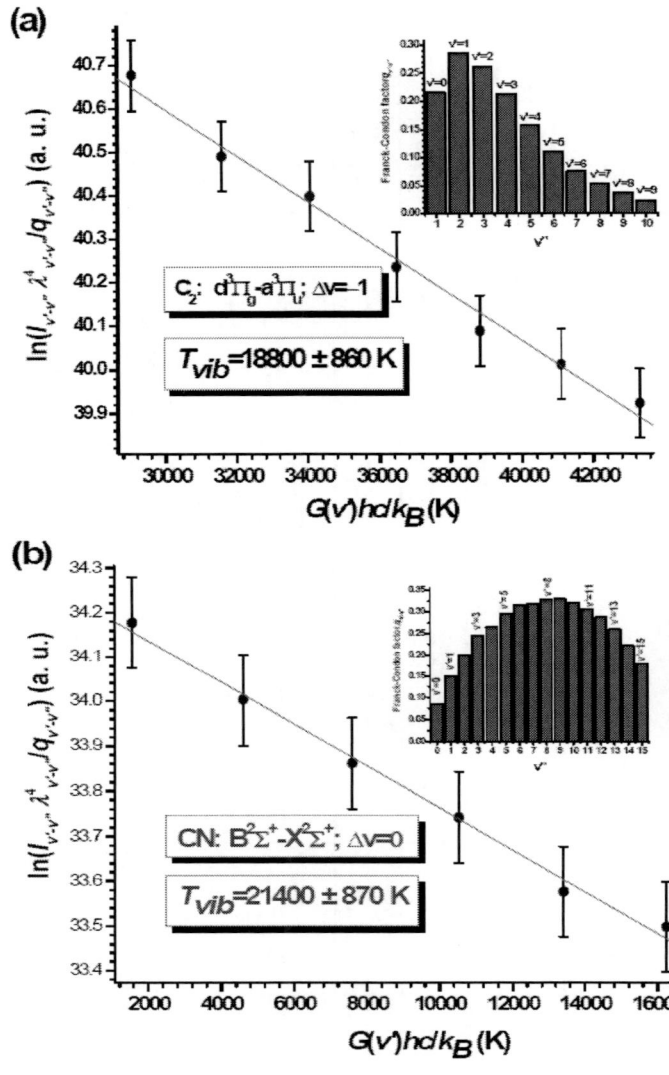

Figure 7. Left (a) panel: Linear Boltzmann plot of the C_2 Swan $\Delta v=-1$ band sequence intensity versus the normalized energy of the upper vibrational level; Right (b) panel: Linear Boltzmann plot of the CN violet $\Delta v=0$ band sequence intensity versus the normalized energy of the upper vibrational level; Experimental conditions: laser power density of 4.5 GW cm^{-2} and vacuum pressure 4 Pa. Plots also show linear fit to the data and the corresponding Franck-Condon factors.

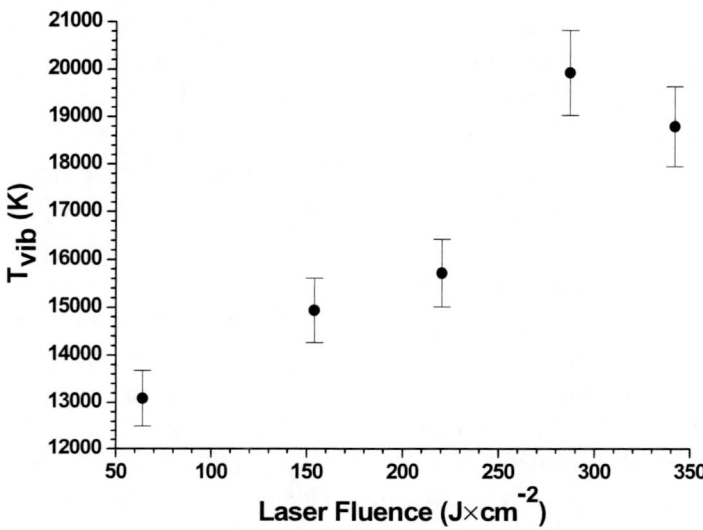

Figure 8. The vibrational temperature T_{vib} calculated from the C_2 Swan $\Delta v=-1$ sequence bands as a function of the CO_2 laser fluence.

The detection of the $C_2(d-a)$ Swan and CN(B-X) bands is of particular interest since it provides an estimation of the plasma vibrational temperature. The emission intensities of the C_2 Swan $\Delta v=-1$ and CN $\Delta v=0$ band sequences were analyzed in order to calculate the molecular vibrational temperature T_{vib}. For a plasma in LTE, the intensity of an individual vibrational v'-v'' band $I_{v'\text{-}v''}$ is given by Eq. (2.24). Two Boltzmann plots of the band intensities against the vibrational energy are given in Figure 7, along with the corresponding Franck-Condon factors. For C_2 and CN the estimated vibrational temperatures were T_{vib}=18800 ± 860 K [Figure 7(a)] and 21400 ± 900 K [Figure 7(b)], respectively. Figure 8 shows the variation of vibrational temperature at 4 Pa of air pressure with laser fluence. The vibrational temperature is maximum at a most efficient laser fluence of 287 J cm^{-2}. These results are consistent with earlier reports on vibrational temperature by different authors [6,11,12,16].

Figure 9 presents a typical resolved LIB emission spectrum of graphite and its rotational assignment for the v'=0-v"=0 band of the $A^2\Delta$-$X^2\Pi$ system of CH. This electronic band is the strongest visible feature of the air, oxygen/acetylene flame and a dominant feature of all hydrocarbon combustion (see figure 4). This spectrum consists in a doublet due to a transition between a Δ upper state (Hund case (b), A_e=-1.11 cm^{-1}) and a Π ground state intermediate between Hund cases (a) and (b) (A_e=+28.1 cm^{-1}) depending on the J value. The fine-structure components are indicated on the branch designations by subscript: $1 \equiv {}^2\Delta_{5/2}\text{-}{}^2\Pi_{3/2}$ and subscript $2 \equiv {}^2\Delta_{3/2}\text{-}{}^2\Pi_{1/2}$. The upper and lower states, Λ doublets are labelled by e and f; when there are the same for both rotational levels ee is abbreviated by e and ff is abbreviated by f. The selection rules involving the parity levels (e or f) are: $e \,\forall\# f$, $e\, \Pi\Pi\, e$, $f\, \Pi\Pi\, f$ for $\Delta J=\pm 1$ (R and P branches) and $e\, \Pi\Pi\, f$, $e\, \forall\# e$, $f\, \forall\# f$ for $\Delta J=0$ (Q branches) [42]. For each value of the quantum number N ($N=J$-S), there are four nearly degenerate energy levels, e or f, $J=N\pm 1/2$. Based on these assumptions we can expect 12 main branches ($\Delta J=\Delta N$ corresponding to R, P and Q branches) and 12 satellite branches ($\Delta J \neq \Delta N$). A partial overlapping of the 0-0 band in the region of P_1 and P_2 branches of the $\Delta v=+2$ of the C_2 Swan band, whereas two lines of C^+ and C^{2+} are also present. To estimate the effective rotational temperature, we consider the J value for the maximum of the 0-0 band (A-X) of CH. This effective rotational temperature is found to be $T_{rot}=2060 \pm 50$ K for $J_{max}=13/2$ of the R_1 branch. Figure 10 shows the simulated spectra for the 0-0 $A^2\Delta\text{-}X^2\Pi$ band of CH calculated (Eq. 2.25) at different temperatures. A good agreement between simulated and observed spectra at a temperature of about 2000 K over the entire range 4200-4400 Å proves that self-absorption is negligible and the rotational levels follow a Boltzmann distribution.

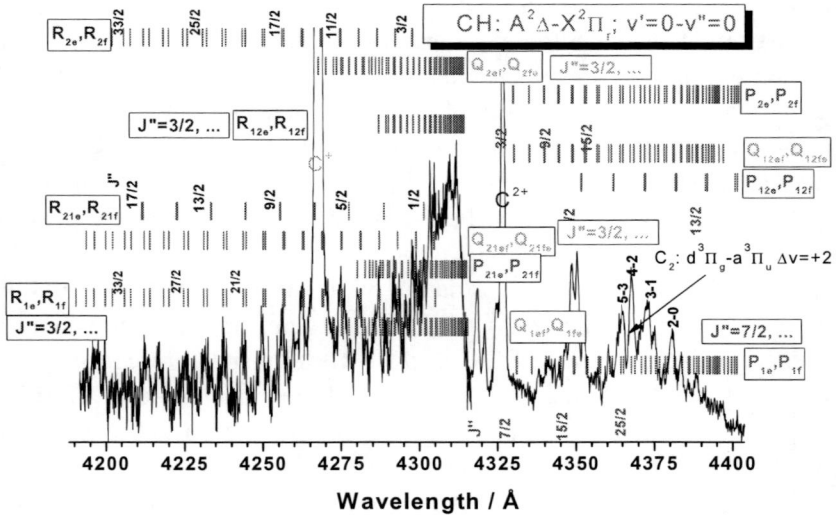

Figure 9. High-resolution PLA of carbon emission spectra at an air pressure of 4 Pa, excited by the 10P(20) line of the CO_2 laser with a laser intensity of 6.31 GW cm^{-2}, and assignment of some ionic carbon lines, the band heads of $C_2(d\text{-}a)$ $\Delta v=+2$ sequence and the rotational structure of the 0-0 $A^2\Delta\text{-}X^2\Pi$ band of CH.

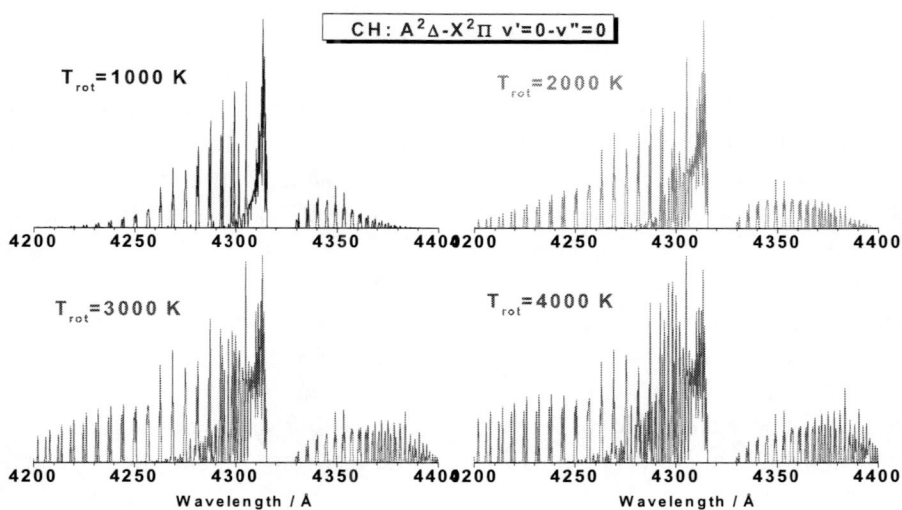

Figure 10. Simulated spectra for the 0-0 $A^2\Delta$-$X^2\Pi$ band of CH calculated at different temperatures.

4.3. IONIZATION DEGREE OF THE PLASMA

As the CO_2 laser beam is focused on the graphite surface, the carbon material absorbs the laser energy to melt, vaporize, and excite the target material. The carbon vapor absorbs more energy and forms high temperature plasma near the surface. The plasma expands into the low-vacuum atmosphere (N_2, O_2, H_2O, etc) and transfers its energy to it. If the pressure around the target is bigger than ~1000 Pa, the breakdown of the air takes place in a significant way. Neutral, single and highly ionized carbon emission lines are found close to the target graphite surface. The carbon clusters and the molecules of the atmosphere obtain an energy that exceeds the binding energy. In these conditions the plasma becomes a mixture of electrons, positive ions such as C^+, C^{2+}, C^{3+}, C^{3+}, C^{4+}, neutral atoms such as C, N, O and H, and molecules such as C_2, CN, CH, NH, and OH in excited electronic states. Let us consider the first three different ionization equilibria of carbon:

$$C(2s^2 2p^2\ {}^3P_0) \leftrightarrow C^+(2s^2 2p\ {}^2P^0_{1/2}) + e + \text{IP(C-I)},$$

$$C^+(2s^2 2p\ {}^2P^0_{1/2}) \leftrightarrow C^{2+}(2s^2\ {}^1S_0) + e + \text{IP(C-II)},$$

$$C^{2+}(2s^2\ ^1S_0) \leftrightarrow C^{3+}(2s^1\ ^2S_{1/2}) + e + IP(C\text{-}III),$$

where the first three ionization potentials (IPs) for carbon are IP(C-I)=11.2603 eV, IP(C-II)=24.3833 eV and IP(C-III)=47.8878 eV [51]. Taking into account the consideration of section 2.4.5, we can obtain the ionization degree. Figure 11 shows the ionization degree $N_i/(N_0+N_i)$ of C, C^+ and C^{2+}, plotted as a function of the gas temperature T at a constant total pressure $P=(N_0+n_e+N_i)k_BT$. The graph shows that carbon is already fully ionized at thermal energies well below the first ionization-energy of 11.2603 eV (equivalent to 130670 K). If we consider a temperature of 23000 K, the ionization degrees of C, C^+ and C^{2+} obtained by means of the Saha equation are 0.999, 0.999 and 0.28, respectively. These so high values of the ionization degrees justify the observed emission spectra. Keeping in mind these results, the temperature obtained from the relative intensity of C^+ lines was chosen as the first approximation for the average excitation temperature.

4.4. ELECTRON NUMBER DENSITY

The electron number density was obtained by considering the discussion reported in section 2.4.3. In our experiments, for C^+ lines, the Doppler line widths are 0.08-0.13 Å at 23000 K (Eq. 2.18). Stark line broadening from collisions of charged species is the primary mechanism influencing the emission spectra in these experiments. In our case, the estimation of electron density n_e has been carried out by measuring the broadening of the spectral profiles of isolated lines of C^+ (2174, 2747, 2837, 2993, 3877, 3920, 4267, and 5890 Å) from the time-integrated high-resolution spectra. The electron number densities of the laser-induced plasma were determined at a laser power density of I_W=1 GW cm^{-2} and air pressure of 4 Pa. A Lorentz functions were used to fit the spectra. In order to extract the Stark broadening from the total experimentally measured line broadening, we have to previously deconvolute the main effects that contribute to the broadening of the spectral line. Values of the electron impact half-width W were taken from the extensive tables given by Griem [27]. Electron densities in the range $(0.69\text{-}5.6)\times10^{16}$ cm^{-3}, with an estimated uncertainty of 10%, were determined. At the evaluated temperature of 23000 ± 1900 K, Eq. (2.4) yields $n_e \approx (0.39\text{-}2.2)\times10^{16}$ cm^{-3}. These electron densities are close to measured values. Based on these calculations, it is difficult to tell whether the plasma is in LTE or not. A possible reason for non-thermal equilibrium could be the large integration time used in the experiments.

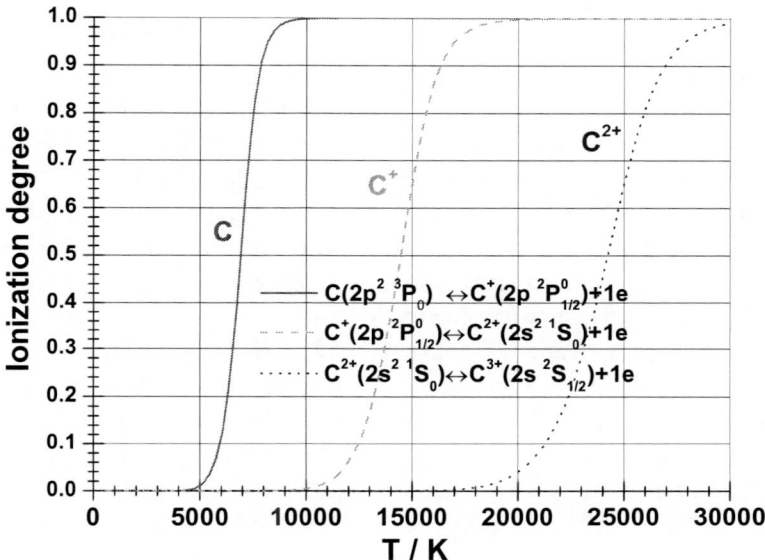

Figure 11. Temperature dependence of the ionization degree $N_i/(N_0 + N_i)$ of carbon C, carbon singly ionized C^+ and carbon doubly ionized C^{2+} at a constant pressure of 4 Pa.

The formation of ionic species is a usual phenomenon in LIB technique. The interaction between the laser and the ablation plume is governed by EII and/or by multiphoton ionization, both followed by electron cascade. EII is the most important for the longer wavelengths used in this work. MPI (Eq. 2.10) on the other hand is relatively improbable for carbon atoms in the ground state $C(2s^2 2p^2\ ^3P_0)$, since its high ionisation potential (11.2603 eV [51]), means that 88 photons are required for this process. The observed ionic emissions are best explained by an EII mechanism (Eq. 2.11). The free or quasifree electrons are produced by the high-power laser pulse at the target surface. These electrons gain sufficient energy from the laser field through inverse bremsstrahlung collisions with neutrals, to ionize carbon atoms or ions by inelastic electron-particle collisions resulting in two electrons of lower energy being available to start the process again (Eq. 2.11). In general, the probability of MPI is $W_{MPI} \propto \Phi_W^n \propto F_E^{2n}$. Calculations of PMI probability for carbon give a negligible value of W_{MPI} for the CO_2 laser at $\lambda=10.591$ μm and $I_W=6.31$ GW×cm^{-2} ($n=88$) (Eq. 2.9).

4.5. EFFECT OF LASER IRRADIANCE

Laser-sample and laser-plasma interactions are strongly dependent on the laser beam irradiance on the target. To see the effect laser irradiance the measurements were also carried out at different laser fluences. Optical emission spectra of the carbon plasma plume in medium-vacuum (~4 Pa) as a function of the laser intensity are shown in Figs. 12(a) and 12(b). These spectra were recorded at a constant distance of 1.5 cm from the target surface along the plasma expansion direction. An increase of atomic and molecular emission intensity with increasing the laser fluence was observed. Figure 13(a) shows the emission intensity change of C(247.856 nm), C^+(251.206 nm), C^{2+}(269.775 nm), C^{3+}(252.998 nm), C^{4+}(227.792 nm), OH 0-0 band head (306.35 nm), and NH 1-0 band head (336.00 nm) as a function of the carbon dioxide laser fluence. The C^{3+}, C^+ and C emission intensity increases drastically with the laser fluence. Beyond ~100 J cm^{-2}, a sharp increase of atomic (especially for C^{3+}, C^+ and C) and molecular line intensities was observed. The C^{2+}, C^{4+}, OH 0-0 band head, and NH 1-0 band head emission intensity increases lightly with the laser fluence. Figure 13(b) shows the emission intensity change of C^+(392.07 nm), C^{2+}(418.69 nm), NH 1-0 band head of the A-X system, CN 1-0, 2-1, 0-0, 1-1 band heads of the B-X violet system, C_2 1-0 band head of the $C^1\Pi_g$-$A^1\Pi_u$ Deslandres-d'Azambuja system, CH 0-0 band head of the *A-X* system, C_2 1-0 band head of the *d-a* Swan system, and H_β line as a function of the laser fluence. An increase of atomic and molecular emission intensity with increasing the laser fluence was observed. Also the background increases with the laser power. At higher laser fluences (154-342 J cm^{-2}), the spectral lines and molecular bands are considerably more broadened than at lower fluences as a result of the high pressure associated with the plasma. It is assumed that at higher laser fluence the PLA plasma is more energetic and more ionized so that the surrounding air can better confine the plasma; the plasma also cools down more rapidly due to the confinement.

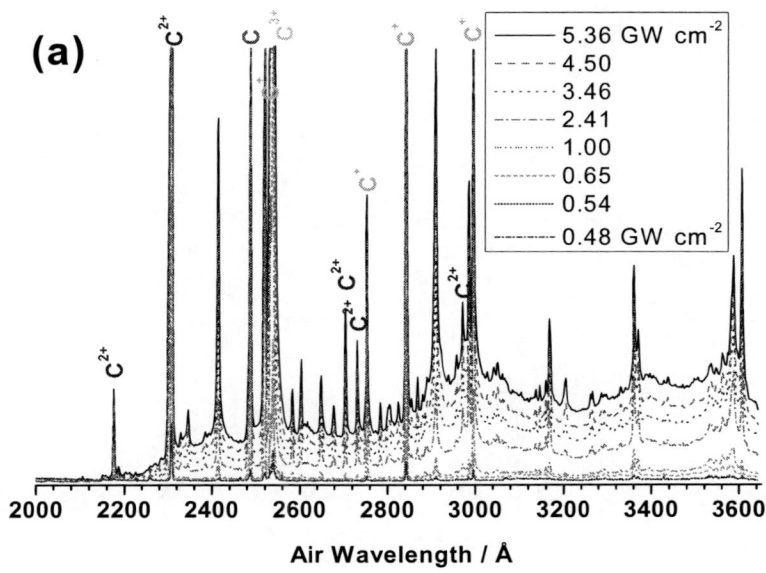
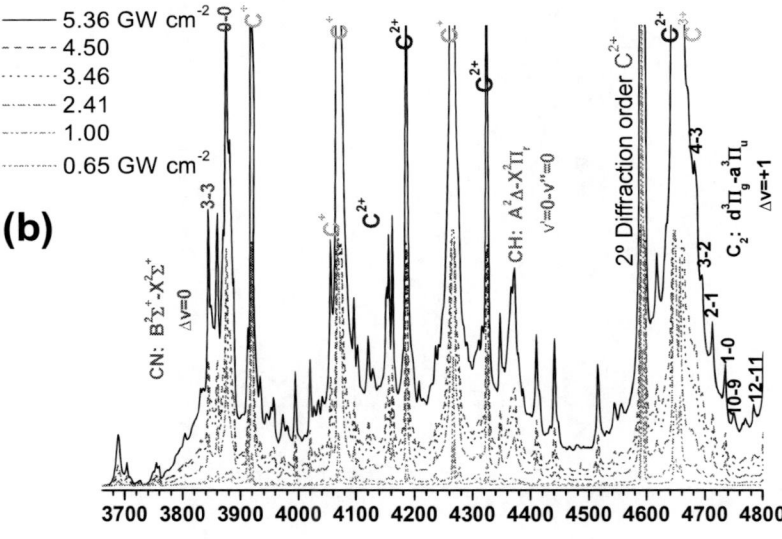

Figure 12. Low-resolution PLA of carbon emission spectrum observed in the (a) 2000-3640 Å and (b) 3660-4800 Å regions, at an air pressure of 4 Pa, excited by the 9P(28) line at 1039.36 cm^{-1} of the CO_2 laser, as a function of the laser power density.

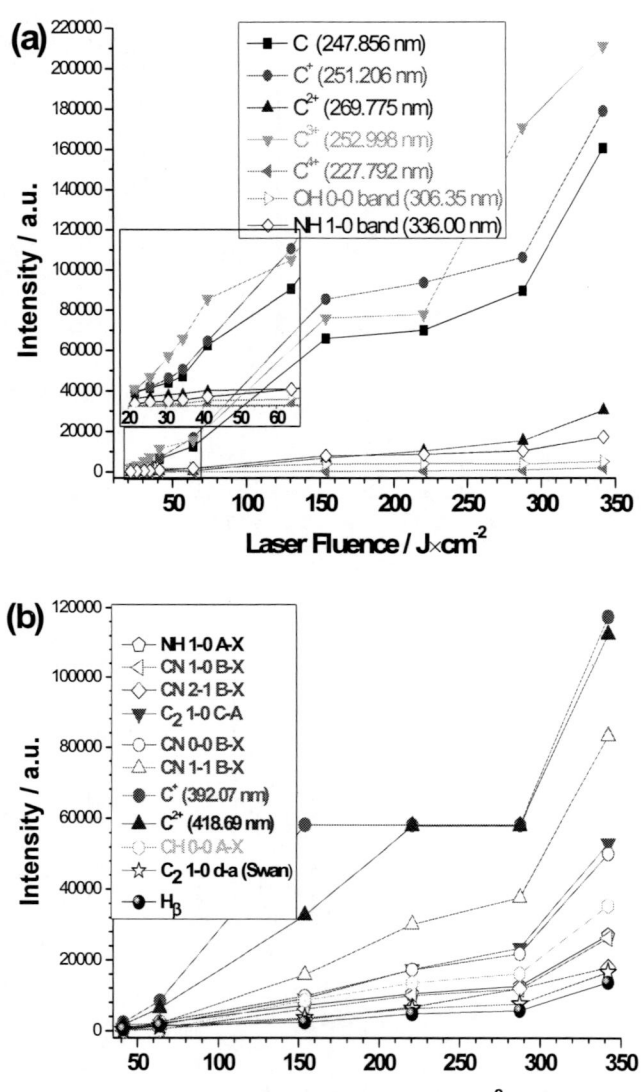

Figure 13 (a)-(b). Emission intensity change of: (a) C(247.856 nm), C^+(251.206 nm), C^{2+}(269.775 nm), C^{3+}(252.998 nm), C^{4+}(227.792 nm), OH 0-0 band head (306.35 nm), and NH 1-0 band head (336.00 nm); (b) C^+(392.07 nm), C^{2+}(418.69 nm), NH 1-0 band head of the $A^3\Pi$-$X^3\Sigma^-$ system, CN 1-0, 2-1, 0-0, 1-1 band heads of the $B^2\Sigma^+$-$X^2\Sigma^+$ violet system, C_2 1-0 band head of the $C^1\Pi_g$-$A^1\Pi_u$, CH 0-0 band head of the $A^2\Delta$-$X^2\Pi$, C_2 1-0 band head of the Swan system $d^3\Pi_g$-$a^3\Pi_u$, and H_β line as a function of the carbon dioxide laser fluence.

4.6. EFFECT OF AMBIENT PRESSURE ON THE PLASMA

The emission characteristics of the laser-induced plasma are influenced to a large extent by the nature and composition of the surrounding atmosphere. The pressure of the air ambient atmosphere is one of the controlling parameters of the plasma characteristics, as well as the factors related to the laser energy absorption. An interesting observation was the effect of the air pressure, studied in the range 4.6 to 63500 Pa. Figs. 14(a)-(b) show typical OES from a carbon plasma plume at different air pressures. These plasma plumes were generated by the CO_2 laser intensity of 1.00 GW cm^{-2}. In general, the spectra of the PLA plume at low pressures (P<1500 Pa) are dominated by emission of electronic relaxation of excited atomic C, N, H, O, ionic fragments C^+, C^{2+} C^{3+} and C^{4+}, and molecular features of $C_2(E-A)$, $C_2(D-X)$, $C_2(d-a)$, $C_2(D-X)$, $C_2(e-a)$, $C_2(C-A)$, $C_2(A-X)$, CN$(D-A)$, CN$(B-X)$, CN$(A-X)$, OH$(A-X)$, NH$(A-X)$, CH$(C-X)$, CH$(B-X)$ and CH$(A-X)$. The spectra of the PLA plume at high pressures (P>10000 Pa) are dominated by emission of electronic relaxation of excited atomic N, O, H, ionic fragments N^+ and O^+, and molecular features of CN$(B-X)$ and CN$(A-X)$. The intensities of the C_2 1-0 band head of the $C^1\Pi_g$-$A^1\Pi_u$ (3607 Å), C_2 1-0 band head of the $D^1\Sigma^+_u$-$X^1\Sigma^+_g$ (4737 Å), CN 1-0 and 0-0 band heads of the $B^2\Sigma^+$-$X^2\Sigma^+$ violet system, CH 0-0 band head of the $A^2\Delta$-$X^2\Pi$ (4307 Å), C^+(3919 Å), C^+(4267 Å), C^{2+}(4593 Å), C^{3+}(4657 Å), and H$_\beta$ spectral lines increase with increasing pressure, reach a maximum at about 200 Pa, and then decrease with higher pressures. Similar results were reported in the literature [52, 53]. Figure 15 shows the evolution of the emission intensity of C_2 1-0 band head of the $C^1\Pi_g$-$A^1\Pi_u$ (3607 Å), C_2 1-0 band head of the $D^1\Sigma^+_u$-$X^1\Sigma^+_g$ (4737 Å), CN 1-0 and 0-0 band heads of the $B^2\Sigma^+$-$X^2\Sigma^+$ violet system, CH 0-0 band head of the $A^2\Delta$-$X^2\Pi$ (4307 Å), N^+(3437 Å), O^+(3410 Å), C^+(3919 Å), C^+(4267 Å), C^{2+}(4593 Å), C^{3+}(4657 Å), and H$_\beta$ atomic lines as a function of air pressure. From figure 15, the intensity of the C^{2+}(4593 Å), C^+(4267 Å), and C^{3+}(4657 Å) spectral lines is found to be more sensitive to the pressure that the CH 0-0 band head of the A–X (4307 Å), C^+(3919 Å), and H$_\beta$ atomic lines. The lines N^+(3437 Å) and O^+(3410 Å) produced in the breakdown of the air, are not observed in the PLA of carbon at low-air pressures. The intensity of CN (Δv=0 sequence) increases with increasing air pressure, reach a maximum at about 200 Pa, and then stays constant as the pressure is increased further. Beyond 200 Pa (see Figure 15), a decrease in the time-integrated emission intensities of C^+, C^{2+}, C^{3+}, CN, C_2, CH, NH and H was found. However, an increase in the emission intensities of the N^+ and O^+ lines was observed. We suggest that these effects are related to shielding by the air plasma, where a part of the laser energy is absorbed by the air plasma during its expansion.

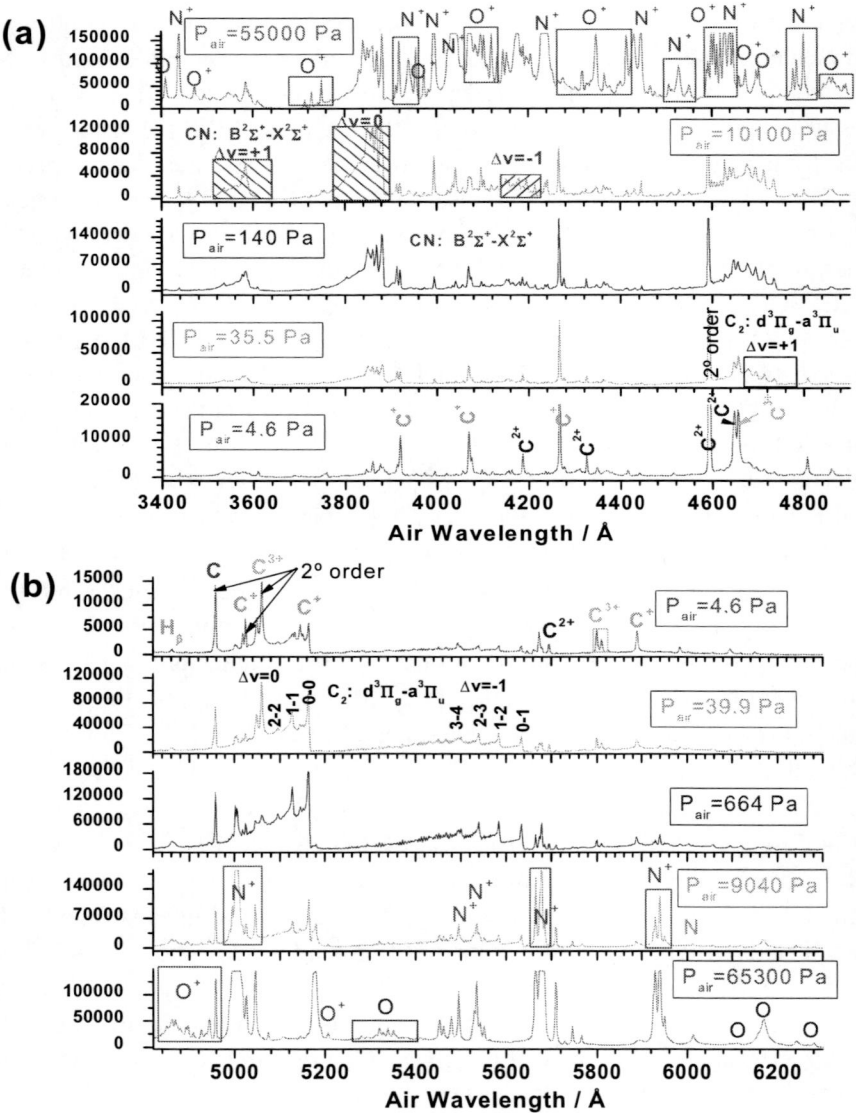

Figure 14 (a)-(b). Low-resolution PLA of carbon OES observed at various air pressures in: **(a)** 3400-4880 Å region and **(b)** 4800-6300 Å region.

This result in a reduction of the atomic and ionic emission intensity of species formed from the carbon target. At low pressures (P<200 Pa), the C, C^+, C^{2+}, C^{3+}, and CN, C_2, CH, NH emissions are produced nearer to the carbon target than the N^+ and O^+ emissions produced nearer to the air plasma position. In general, the air ambient gas will confine the plasma near the target (produced mainly by C, C^+, C^{2+}, C^{3+}, and CN, C_2, CH) and prevent the electrons and species produced near the target escaping quickly from the laser focal volume (observation region). Therefore, the emission intensity increases with increasing pressure. However, at higher pressures (more than 200 Pa in our case), the ambient gas will hinder the plasma from penetrating the atmosphere and predictably cause a higher plasma temperature. The emission intensity of H, C, C^+, C^{2+}, C^{3+}, and CN, C_2, CH decreases because of the fact that the laser energy is absorbed by air, producing air breakdown and increasing the N, O, N^+, and O^+ emission intensity, in agreement with our observation in Figure 15. At lower air pressures, the absence of the shielding air plasma results in a strong increase in the intensity of the C, C^+, C^{2+}, C^{3+} emission from the carbon target plasma. At such lower air pressures the relative contribution of the N^+, O and O^+ emission diminishes, and the emission from carbon surface component becomes dominant.

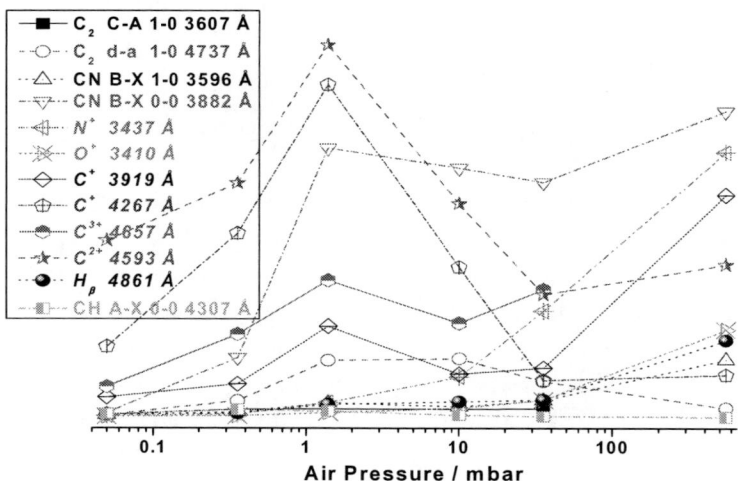

Figure 15. Emission intensity change of C_2 1-0 band head of the $C^1\Pi_g$-$A^1\Pi_u$ (3607 Å), C_2 1-0 band head of the $d^3\Pi_g$-$a^3\Pi_u$ (4737 Å), CN 1-0 and 0-0 band heads of the $B^2\Sigma^+$-$X^2\Sigma^+$ violet system, CH 0-0 band head of the $A^2\Delta$-$X^2\Pi$ (4307 Å), N^+ (3437 Å), O^+ (3410 Å), C^+(3919 Å), C^+(4267 Å), C^{2+}(4593 Å), C^{3+}(4657 Å), and H_β line as a function of the air pressure around the carbon target.

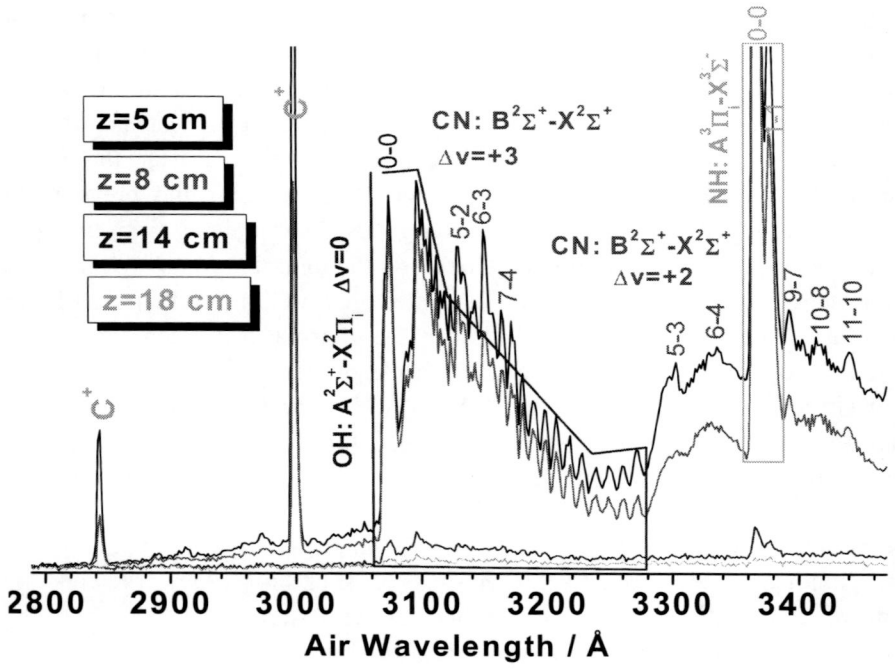

Figure 16. Optical emission spectra of the graphite ablation plume at: 5, 8, 14, and 18 cm from the target.

4.7. SPATIAL CHARACTERIZATION

In this section we present experimental results on the laser ablation of a graphite target at P_{air}=4 Pa by using a high-power IR CO_2 pulsed laser (λ=9.621 μm and laser fluence of 342 J cm^{-2}) and distances from 0.2 up to 20 cm, from the target along the plasma expansion direction. We discuss the dynamics of the plume expansion and formation of different atomic (C, N, H and O), ionic (C^+, C^{2+}, C^{3+} and C^{4+}) and molecular (C_2, OH, CN, CH and NH) species. Although OES gives only partial information about the plasma particles, this diagnostic technique helped us to draw a picture of the plasma in terms of the emitting chemical species, to evaluate their possible mechanisms of excitation and formation and to study the role of gas-phase reactions in the plasma expansion process, allowing a discussion of the probable OH, NH, CH, CN, H, O and N formation by gas phase reactions during the propagation of the plasma plume.

Results and Discussion

The plasma emission was recorded at several distances along the plasma expansion direction (Z axis of Figure 2) at a constant distance $y=10$ cm with respect to the focusing lens onto the entrance slit of the monochomator and parallel to the target surface. Figures 16-18 show time-integrated OES following nanosecond pulsed laser ablation of graphite monitored at several distances from the target in different spectral regions. One can see from these figures that the emission along the plasma Z-axis is, in the conditions of sensitivity of our detection system, about 20 cm. Also we can observe that the intensity of the C^{2+}, C^{3+} and C^{4+} ionic emission lines decays rapidly at distances higher to 1.5 cm. On the other hand it can be see that at high distances from the graphite target ($z=18$ cm) atomic lines from C^+, C, H, O, and N are still observed. The intensities of the molecular bands for different species (OH, NH, CH, CN and C_2) generally decay with distance, being observed up to 18 cm.

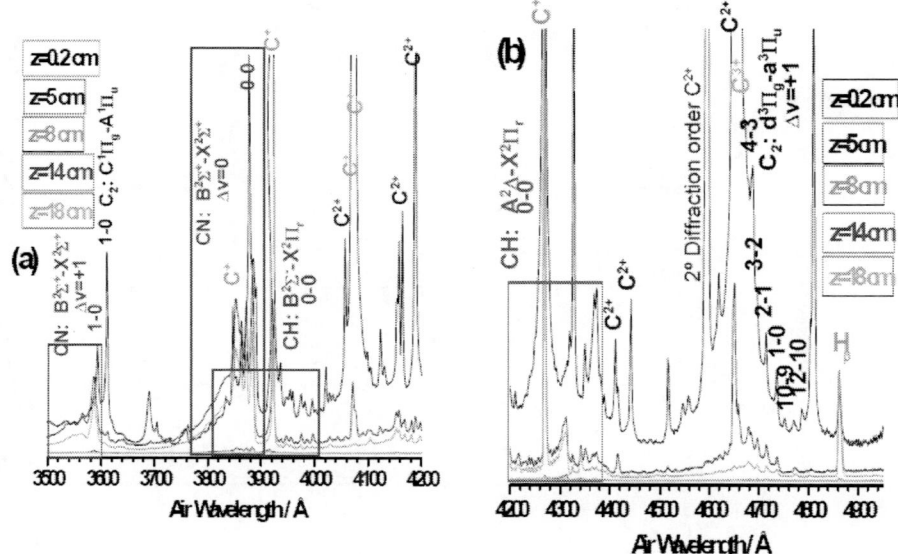

Figure 17. Optical emission spectra of the graphite ablation plume monitored at: 0.2, 5, 8, 14, and 18 cm from the target: (a) 3500-4200 Å; (b) 4200-4950 Å regions.

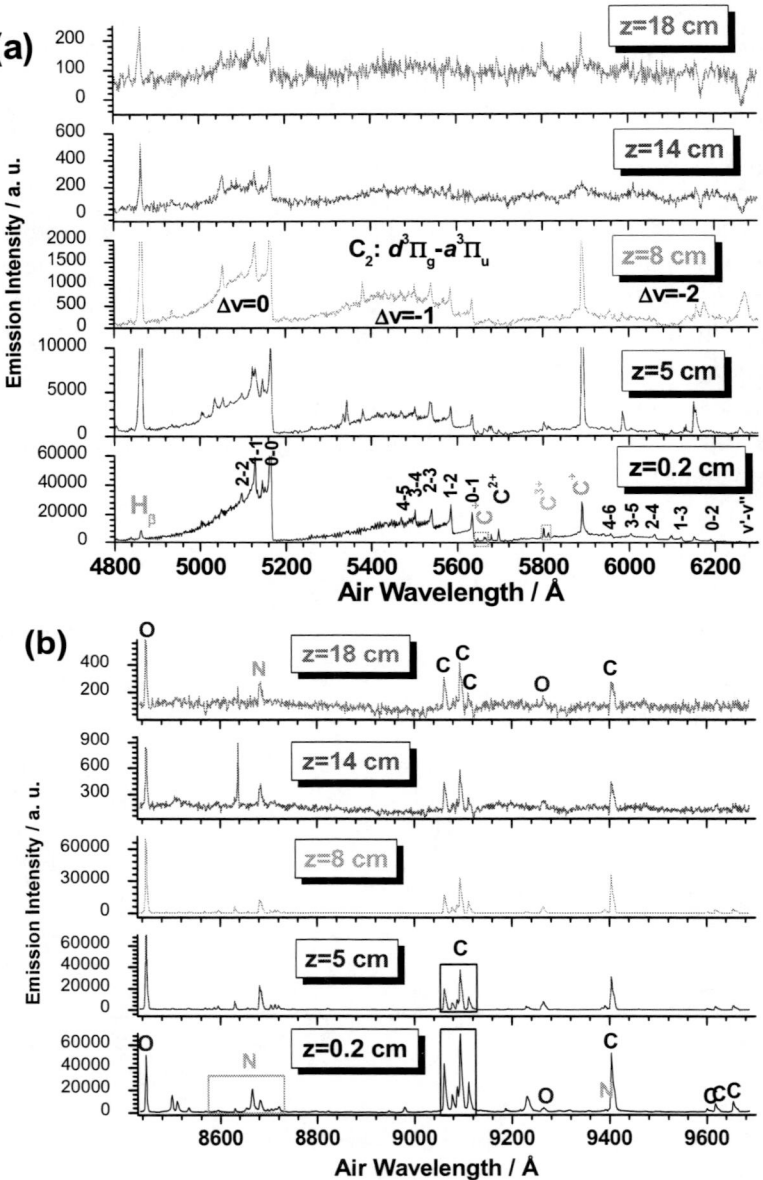

Figure 18 (a)-(b). Optical emission spectra of the graphite ablation plume monitored at: 0.2, 5, 8, 14, and 18 cm from the target in the spectral region: (a) 4800-6250 Å and (b) 8430-9700 Å.

Figure 19(a) shows the emission intensity change of C^+(3920.7 Å), C^{2+}(4186.9 Å), H_β(4861.36 Å), NH 1-0 band head (3365 Å), CN B-X 1-0, 0-0, and 1-1 band heads (3591, 3887, 3875 Å, respectively), C_2 C-A 1-0 band head (3615 Å), CH A-X 0-0 band head (4317 Å), and C_2 d-a 1-0 band head (4743 Å), as a function of the distance. The C^{2+} emissions are only observable close to the carbon target (z<4 cm). The intensity of CN, NH, CH bands, and H_β line increase lightly with increasing the distance, reach a maximum at about 5 cm, and then stay constant as the distance is further increased. Beyond 8 cm, a decrease in the time-integrated emission intensities of these species was found. The intensity of the C_2 band falls continuously with the distance to the target. The continuous decrease of the C^+, C^{2+}, and C_2 emission intensities with the distance indicates that these species are mainly formed from the carbon surface. This decrease is more drastic for the C^{2+} emission intensity probably due to the fact that the recombination with free electron is faster than for C^+ species. The slight rise of the NH, CN, CH and H_β emission intensities as a function of the distance up to $z_{max} \approx 5$ cm indicates that these electronically excited species are mainly formed by gas phase reactions during the propagation of the plasma plume through the air gas. The NH, CN, CH and H_β emission intensities diminish for z > 8 cm due to the decrease in density during the expansion of the plasma plume. In figure 19(b), the emission intensity change of O(8446.359 Å), O(9262.776 Å), N(8680.28 Å), C(9061.43 Å), C(9078.28 Å), C(9088.51 Å), C(9094.83 Å), C(9111.80 Å), and CN A-X 1-0 band head at 9192 Å as a function of the distance from the target are shown. The intensity of O(8446.359 Å), O(9262.776 Å), N(8680.28 Å) and CN A-X 1-0 band head increases lightly with increasing the distance from the target expansion direction, reach a maximum at about 5 cm, and then stays constant as the distance is further increased. Beyond 8 cm a decrease in the time-integrated emission intensities of these species was found. However, all the carbon emission lines decrease with increasing the distance from the target. As previously discussed in figure 19(a) the behaviour of carbon species are mainly formed from the surface target but, molecular species such as NH, CN and CH are produced in gas phase.

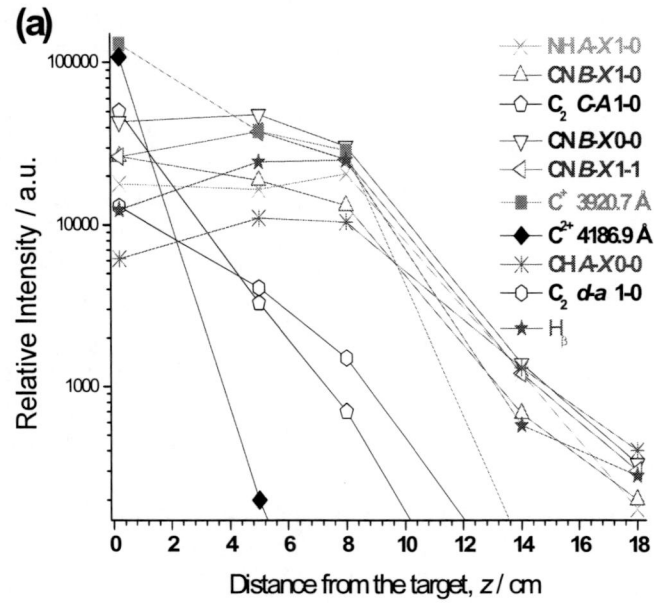
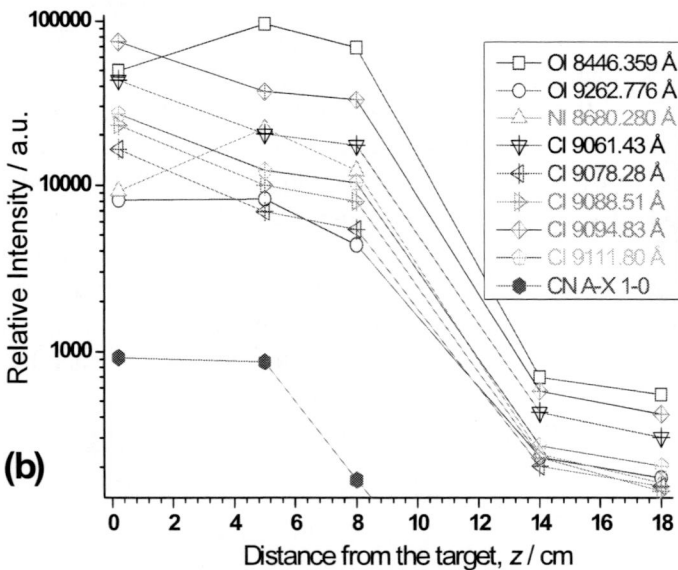

Figure 19. (a)-(b). Emission intensity change of: **(a)** C^+, C^{2+}, H_β, NH, CN, C_2 (*C-A* 1-0 band head), CH, and C_2 (*d-a* 1-0 band head) and **(b)** O (two lines), N, C (five lines) and CN (*A-X* 1-0 band head), as a function of the *z* distance.

In order to further identify properties of the ablation plasma plumes originated from graphite targets, we have estimated the vibrational temperatures (Eq. 2.24) of C_2 molecule (*d-a* Swan $\Delta v=-1$ band sequence) as function of *z*-distance. The estimated vibrational temperatures were T_{vib}=23000±1000, 16200±900, 10800±600, 7700±500 K at 0.2, 1.5, 5 and 9 cm from the target along the plasma expansion direction, respectively compatible with a cooling stage.

4.8. TEMPORAL EVOLUTION OF THE PLASMA

In this section time-resolved OES analysis for the plasma plume, produced by high-power tunable IR CO_2 pulsed laser ablation of graphite, at λ=10.591 µm and a laser fluence of 402 J cm^{-2} is presented. We focus our attention on the temporal evolution of different atomic/ionic and molecular species over a broad spectral range from 190 to 1000 nm. Excitation temperature, electron density and vibrational temperature in the laser-induced plasma were estimated from the analysis of spectral data at various times from the laser pulse incidence.

In time-resolved measurements, the delay t_d and width t_w times were varied. It was verified that the plasma was reproducible over more than 7 ablation events by recording the same spectrum several times. The temporal history of laser-induced breakdown carbon plasma is illustrated schematically in Figure 20. The time when beginning of the CO_2 laser pulse is triggered is considered as the origin of the time scale (t=0). Inserts illustrate some emission spectra recorded at different delay and width times. The temporal shape of the CO_2 laser pulse is also shown. Because the LIB plasma is a pulsed source the resulting spectrum evolves rapidly in time.

The LIBS spectra of carbon were measured at different delay and width times. As an example figure 21(a) show OES of the graphite ablation plume at low-resolution in 2000-2800 Å region monitored at several delay times for a fixed gate width time of 0.1 µs. At early times (t_d<2 µs) the predominant emitting species are the C^{2+} $2p^2$ $^1D_2 \rightarrow 2s2p^2$ $^1P_1$ atomic line at 2296.87 Å and two lines of C^{3+} at 2524.41 and 2529.98 Å. The three molecular band systems observed in this spectral region are the $C_2(E^1\Sigma_g^+ - A^1\Pi_u$; Freymark), $C_2(D^1\Sigma_u^+ - X^1\Sigma_g^+$; Mulliken) and $C_2(e^3\Pi_g - a^3\Pi_u$; Fox-Herzberg). These bands stay approximately constant as the delay time is increased up to ~5 µs, and decrease for higher delay times.

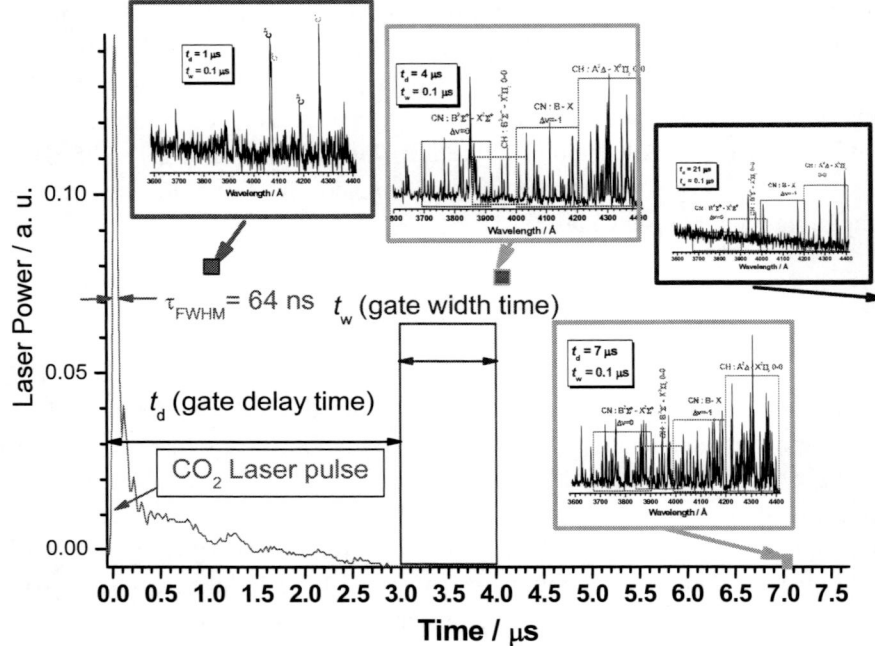

Figure 20. A schematic overview of the temporal history of laser-induced breakdown carbon plasma. Here t_d is the gate delay time and t_w is the gate width time during which the plasma emission is monitored. Inserts illustrate some spectra observed at different delay and width times. The temporal shape of the CO_2 laser pulse (recorded with the aid of the photon-drag detector) is also shown.

Figure 21(b) shows the LIB emission spectrum of graphite plasma plume recorded 1, 1.2, 4 and 11 μs after the CO_2 laser irradiation in the spectral region 2850-3600 Å. The main features in this region are the emission of C^+ ionic species and several molecular emission bands from C_2, OH and CN. The C^+ emission intensities fall considerably as the delay changes from 1 to 11 μs. On the other hand, the molecular bands of C_2, OH and CN emission intensities stay approximately constant as the delay time is increased up to ~5 μs, and decrease for higher delay times. From our results at low-resolution we can appreciated that the intensity of the C^+, C^{2+} and C^{3+} ionic emission lines decays rapidly at delay times higher to 2 μs. On the other hand we can see that at high delay time after plasma ignition, atomic lines from C, H, O, and molecular bands of C_2, CN and OH are still observed.

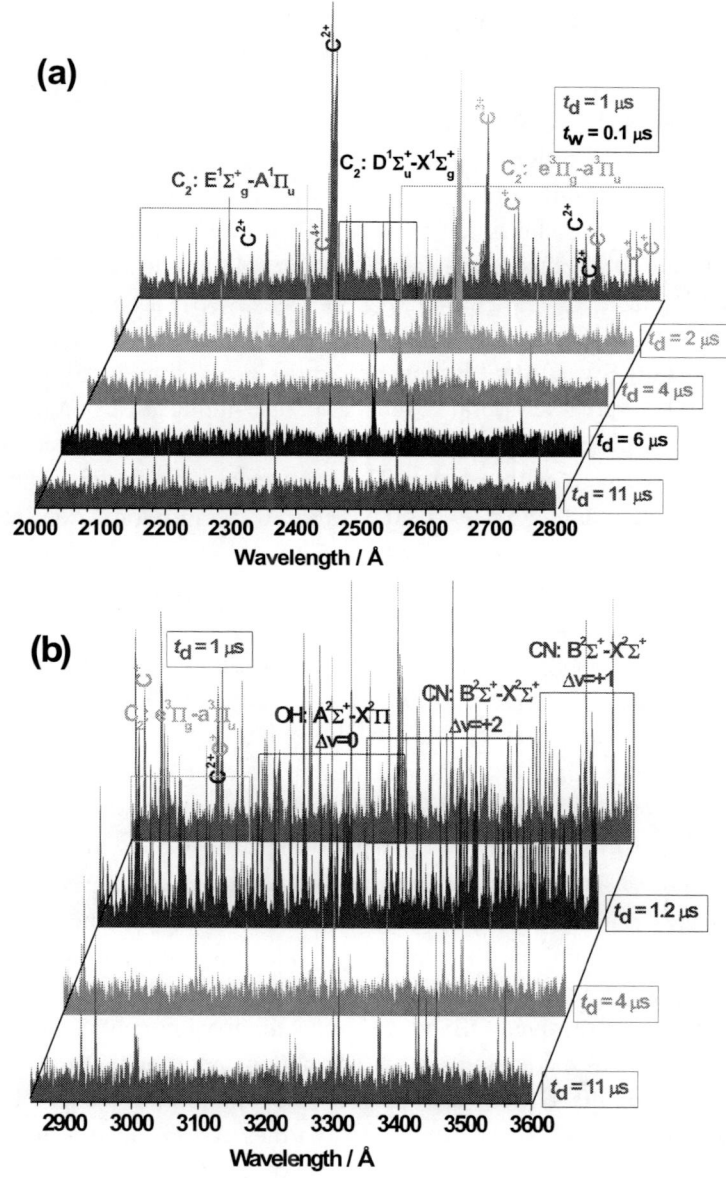

Figure 21. Optical emission spectra of the graphite ablation plume monitored at different delay times for a fixed gate width time of 0.1 µs in two spectral regions.

In order to get more insight into laser ablation of graphite and to understand the laser-induced breakdown dynamics, we have scanned in the UV-Visible spectral region with higher resolution. As an example some results for the spectral region 4640-4750 Å are shown in figures 22-24. The spectral range was chosen in order to detect both double and triple ionized carbon species and C_2 diatomic molecule. In figure 22, the data acquisition was performed by averaging the signal over: (a) 20 successive laser shots ($t_d=0$ and $t_w>>30$ μs) and (b) 7 successive laser shots ($t_d=4$ μs and $t_w=0.02$ μs). The emissions of ionized $C^{2+}(1s^22s3p\ ^3P^0_{2,1,0} \rightarrow 1s^22s3s\ ^3S_1)$ around 4650 Å, and C^{3+} around 4658.3 Å are considerably higher in the spectrum of Figure 22a, while the C_2(d-a) Swan $\Delta v=+1$ sequence emission is similar. Figures 23a-b and 24 show the typical temporal sequence of laser-induced carbon plasma. At early times ($t_d\leq 0.02$ μs) emission from C^{2+} and C^{3+} is easily detected between 4645-4670 Å (see inset within figure 23-a). As seen in figure 23-b during the initial stages after laser pulse ($t_d\leq 0.04$ μs), C^{2+} emissions dominate the spectrum. As time evolves (0.04 μs$\leq t_d\leq 1.5$ μs), C^{3+} emission dominate the spectrum. As the delay is increased up to 2.5 μs (1.5 μs$\leq t_d\leq 2.5$ μs) again C^{2+} emission dominates the spectrum. These ionic lines decrease quickly for higher delay times, being detected up to ~ 3 μs. Some oxygen and nitrogen ionic lines were also observed in the spectra at the gate delay from 0.02 μs to 1 μs and its emission intensities remain approximately constant in this time interval (see Figure 23a). They vanished after the delay of ~1.5 μs. It shows that the air is ionized by the CO_2 laser pulse and by the collisions with the laser induced plasma. During the time period up to ~ 0.5 μs, no apparent C_2 emissions were observed. As can be seen from Figure 24, the C_2(d-a; $\Delta v=1$ band sequence) emissions were clearly observed from ~2 μs. The C_2 emission intensities increase lightly with increasing t_d, reach a maximum at ~5 μs, and then decrease as the time is further increased.

Space-and-time resolved OESs laser-induced measurements could be used to estimate plasma expansion rate. To obtain additional time resolved information about the optical emission of the plume, wavelength resolved spectra have been recorded at different delay times at a distance of 9 mm. The temporal evolution of spectral atomic, ionic and molecular line intensities at a constant distance from the target can be used to construct the time-of-flight (TOF) profile. TOF studies of the emission provide fundamental information regarding the time taken for a particular species to evolve after the laser-induced plasma has formed. Specifically, this technique gives an indication of the velocity of the emitted species. A rough estimation of the velocity for the different species in the plume can be inferred from the time resolved spectra by plotting the intensities of selected emission lines versus the delay time, and then calculating the velocity by dividing the distance from the target by the time where the emission peaks. This method for determination of plasma velocity should be used with care due to the superposition of both expansion and forward movements of the plasma plume.

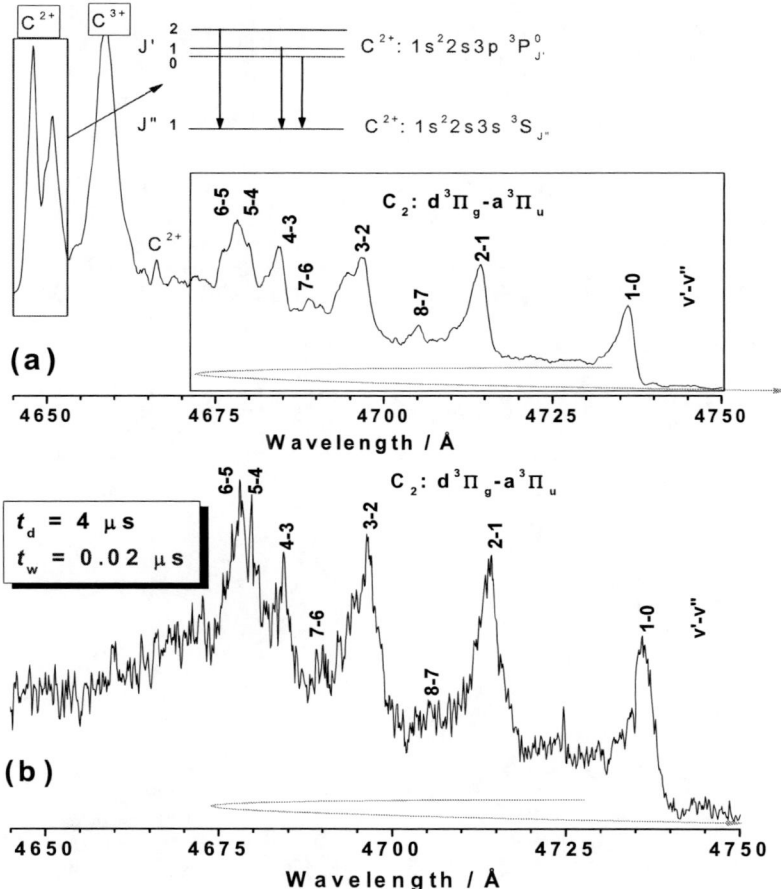

Figure 22. (a)-(b). Measured high-resolution pulsed laser ablation of graphite emission spectra observed in the region 4645-4750 Å region. The data acquisition was performed by averaging the signal over: (a) 20 successive laser shots with $t_d=0$ and $t_w>>30$ μs; (b) 7 successive laser shots with $t_d=4$ μs and $t_w=0.02$ μs. The assignments of some ionic lines of C^{2+} and C^{3+} and molecular bands of C_2 are indicated. The insert in (a) illustrates the rotational structure of one triplet of C^{2+} line.

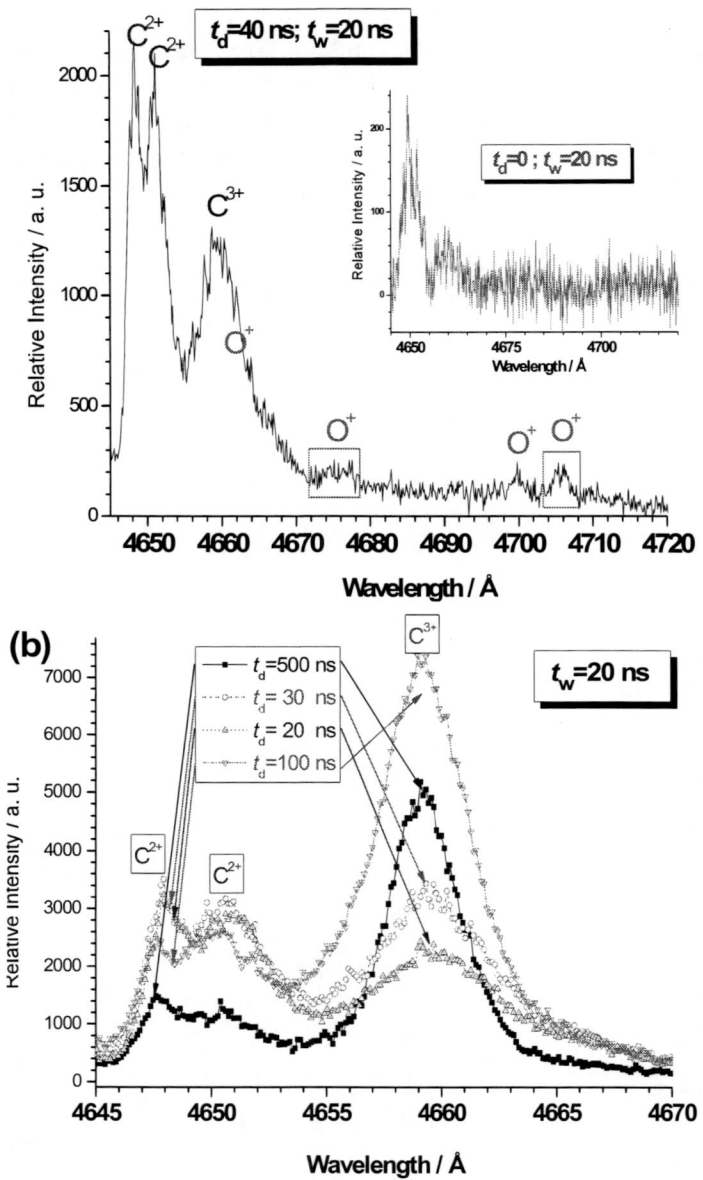

Figure 23.(a)-(b). Time-resolved high-resolution emission spectra from laser-induced carbon plasma observed in the region: **(a)** 4645-4720 Å region monitored at 40 ns delay time; **(b)** 4645-4670 Å region monitored at 20, 30, 100, and 500 ns gate delay times for a fixed gate width time of 20 ns. The inset in (a) displays the spectrum the first 20 ns after incidence of the laser pulse.

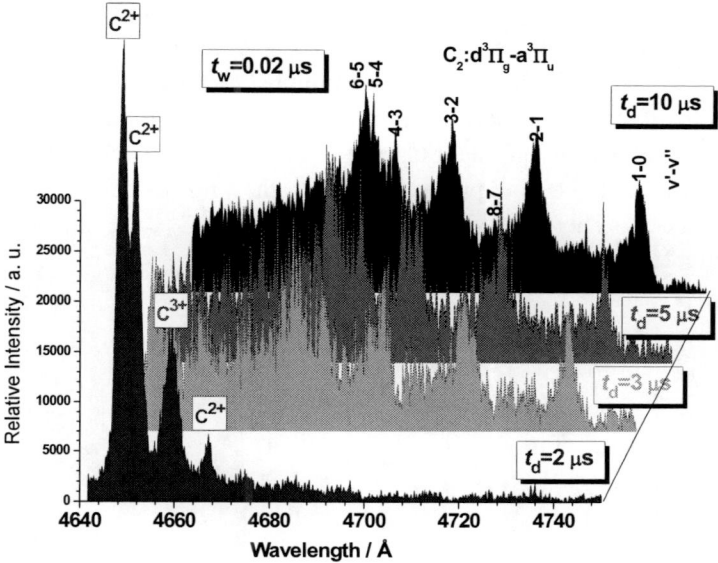

Figure 24. Time-resolved high-resolution emission spectra from laser-induced carbon plasma observed in the region 4645-4750 Å region monitored at 2, 3, 5, and 10 μs gate delays for a fixed gate width time of 20 ns.

Figure 25 displays the TOF profile, for ablation experiments induced by CO_2 laser pulses, of several C, C^+ and C^{3+} lines intensities in UV region and C^{2+} in the visible as a function of delay time. However, the insert of the figure shows the time dependence of C^{2+} and C^{3+} line intensities in the visible region for ablation induced by CO_2 laser pulses in which the tail has been eliminated by means of the suppression of the N_2 in the gas mixture of the active laser medium. All the data are taken from high-resolution spectra and in the figures the temporal profiles of both kinds of laser pulses are also plotted. In both cases emissions from C^{3+} are stronger than emissions coming from the other species. All the ionic lines follow the time profile of lasers pulses lasting until four or three microseconds depending on the kind of the laser pulse. These behaviours may be related to the laser absorption processes on the target surface. Thus for "non tailed pulses" the line intensities start to growth at 400 ns while for "tailed pulses" start at 70 ns. Since the energy pumping (rise time) of each kind of pulse is different, the species reach the maximum intensity at different times: ~1 μs for non tailed pulses and 700 ns for tailed ones, indicating that the graphite target needs some energy threshold to eject the different species. The higher intensity in the 0-400 ns time interval for the C^{2+} may be due to the higher sensibility of our ICCD camera in the visible region than in the UV one. The different behaviour of atomic C can be also observed in figure 25. Atomic C have a higher rise time and lasting more (> 15 μs) than ionic species, possibly due to the continuous recombination of ions with electrons to give excited carbon. From these results has not been observed excitation dependence on the pulse tail however the energy of the pulse intensity seems to be the pulse parameter that influence on the graphite ablation process. The peak velocities estimated for C^{3+}, C^{2+} and C^+ species, from figure 25, are about 7×10^3 m/s.

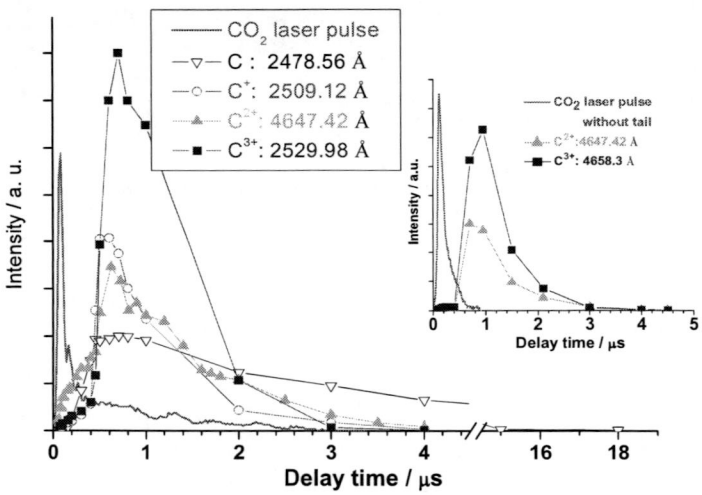

Figure 25. Emission intensity change of C(2478.56 Å), C^+(2509.12 Å), C^{2+}(4647.42 Å) and C^{3+}(2529.98 Å) lines as a function of delay time (fixed gate width time of 20 ns) for a CO_2 pulse laser with a tail of about 3 μs. The insert shows the emission intensity change of C^{2+}(4647.42 Å) and C^{3+}(4658.3 Å) lines as a function of delay time (fixed gate width time of 20 ns) for a CO_2 pulse laser without tail.

In this section the plasma temperature was determined form the emission line intensities of several C^+ lines observed in the laser-induced plasma of carbon target for a delay time of 1 μs and 0.02 μs gate width. The obtained excitation temperature was 26000 ± 3000 K. The carbon ionic multiplet line at ~3920 Å was identified as candidate for electron-density measurements. Figure 26-a shows, the 3920 Å carbon ionic line with sufficient resolution to measure the full width at half-maximum at 8 different time delays. All the data points were fitted with Lorentzian function to determine the Stark line width. By substituting these values in Eqn. (2.21) and the corresponding value of electron impact parameter W (0.465 Å from Griem [27] at plasma temperature of 26000 K), we obtain the electron density. Figure 26-b gives the time evolution of electron density by setting the gate width of the intensifier at 0.02 µs. The initial electron density at 0.02 µs is approximately 3×10^{16} cm^{-3}. Afterwards, the density increases over the period of 0.1 µs and reaches a maximum at 0.1 µs (time period of the peak CO_2 laser pulse), and then decrease as the time is further increased. At shorter delay times (<0.1 µs), the line to continuum ratio is small and the density measurement is sensitive to errors in setting the true continuum level. For times >0.1 µs, the line to continuum ratio is within reasonable limits and the values of electron density shown in figure 26-b should be reliable. Initially the laser-induced plasma expands isothermically within the time of the duration of the laser pulse. After termination of the peak laser pulse (~0.1 µs) the plasma expands adiabatically. During this expansion the thermal energy is converted into kinetic energy and the plasma cools down rapidly. After 4 µs, the electron density is about 1.5×10^{16} cm^{-3}. For a long time >4 µs, subsequent decreased C^+ emission intensities result in poor signal-to-noise ratios, and there exits a limitation in the spectral resolution. The decrease of n_e is mainly due to recombination between electrons and ions in the plasma. These processes correspond to the so-called radiative recombination and three-body recombination processes in which a third body may be either a heavy particle or an electron.

In order to further identify properties of the ablation plasma plumes originated from graphite targets, we have estimated the vibrational temperatures of C_2 molecule as function of delay time. The emission intensities of the C_2 d-a Swan Δv=+1 band sequence were used to estimate these vibrational temperatures T_{vib}. The estimated vibrational temperatures were T_{vib}=8000±500, 8300±600, 7500±600, 4500±900 K at 3, 5, 10 and 15 µs after plasma ignition, respectively compatible with a cooling stage.

Optical emission accompanying TEA-CO_2 nanosecond laser ablation of carbon is very long lived (~40 µs) relative to the average radiative lifetimes of the excited levels that give rise to the observed emission lines. At distances close to the target surface (<9 mm), all of the emission lines of C, C^+, C^{2+} and C^{3+} expected in the 2000-10000 Å wavelength range are observed, illustrating that the excited species giving rise to the optical emission are produced by non-specific mechanism during the TEA CO_2 laser ablation process. However, a direct excitation-de-excitation mechanism cannot explain the observed emission spectra. EII would explain the emission intensity variation with the time for C, C^+, C^{2+}, C^{3+} and C_2 species. On the other hand, the formation of the excited molecular species would happen in gas phase by collisions between atomic or ionic species present in the plume and the residual gas at times far away from the plasma ignition. The emission process at this plasma stage is divided into two different process associated, respectively with the shock formation and the plasma cooling. During the former, the atoms, molecules and ions gushing out from the carbon target are adiabatically compressed against the surrounding gas. During the latter stage the temperature of the plasma and consequently the emission intensities of atomic lines and molecular bands decrease gradually.

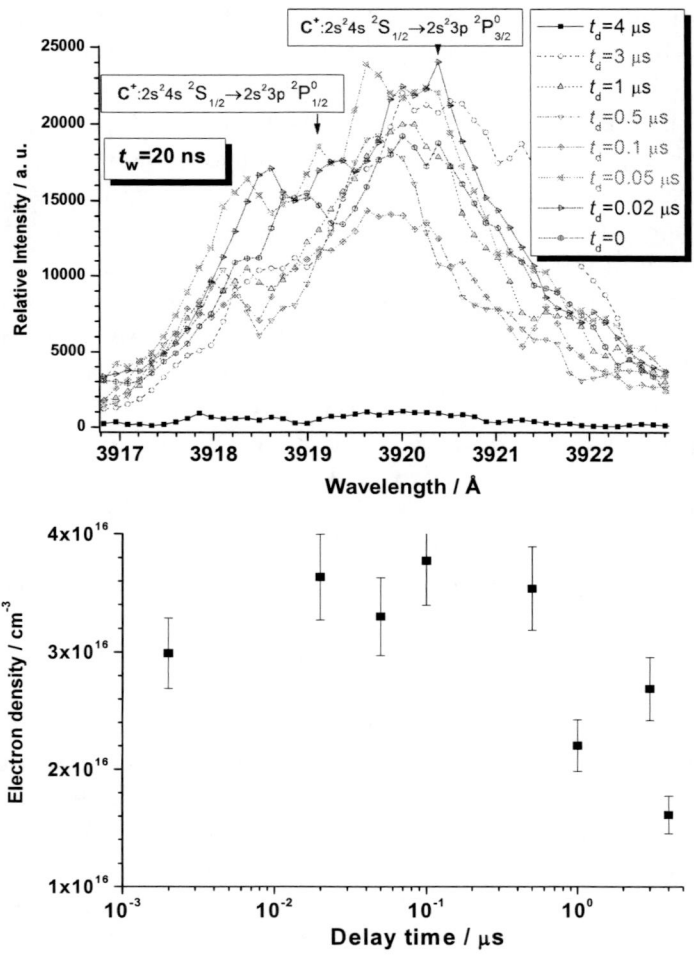

Figure 26. (a). Stark-broadened profiles of the C^+ line at 3920 Å at different delay times for a fixed gate width time of 0.02 μs. (b) Temporal evolution of electron density at different delay times from plasma ignition.

The evolution of the TEA-CO_2 laser-induced carbon plasma can be divided into several transient phases. The initial plasma (t_d<2 µs) is characterized by high electron and ion densities (10^{16}-10^{19} cm^{-3}), and temperatures around 2.2 eV. The emission spectrum from this early stage is characterized by emission lines from C^{3+}, C^{2+} and C^+ ions. Owing to the high electron density, the emission lines are broadened by Stark effect. The ionic emission lines (C^{3+}, C^{2+}, C^+) decay rapidly being observed up to ~3 µs. Emission lines from C atoms and molecular species (C_2, CN, CH, OH) in excited electronic states can be found after about 1 µs time delay. As the plasma expands and cools, the electrons and ions recombine. After the initial plasma (t_d>3 µs), the molecular emissions increase slowly up to ~5 µs and after that decay slowly up to ~40 µs.

Chapter 5

CONCLUSION

This article reviews some fundamentals of LIBS and some experimental studies developed in our laboratory on the ablation of graphite using a high-power IR CO_2 pulsed laser. In this experimental study we used several laser wavelengths (λ=9.621 and 10.591 µm) and laser intensity ranging from 0.22 to 6.31 GW cm^{-2}. Ablation was produced typically at medium-vacuum conditions (~ 4 Pa). Emissions from the resulting ablation plumes, and from the collisions with the ablated material and the background gas molecules (N_2, O_2, H_2O, etc), have been investigated by wavelength-, space-, and time-resolved OES from UV-Vis-NIR. Wavelength-dispersed spectra of the plume reveal C, C^+, C^{2+}, C^{3+}, C^{4+}, N, H, O, N^+, O^+ and molecular features of C_2, CN, OH, CH, N_2, N_2^+ and NH emissions corresponding to different electronic band systems. For the assignment of molecular bands a comparison with conventional emission sources was made. Excitation, vibrational and rotational temperatures, ionization degree and electron number density for some species were estimated by using different spectroscopic methods. The characteristics of the spectral emissions from the different species have been investigated as functions of the ambient pressure, laser irradiance, the distance from the target and delay time after plasma ignition. Time-gated spectroscopic studies have allowed estimation of TOF and propagation velocities for various emission species. Possible production routes for secondary emitters such as C_2, CN, OH, CH, N_2, N_2^+ and NH are discussed.

ACKNOWLEDGMENTS

We gratefully acknowledge the support received in part by the DGICYT (Spain) Projects: MEC: CTQ2007-60177/BQU and MEC: CTQ2008-05393/BQU for this research.

REFERENCES

[1] Kroto, HW; Heath, JR; O'Brien, SC; Curl, RF; Smalley, RE. *Nature,* 1985, 318, 162-163.
[2] Koinuma, H; Kim, MS; Asakwa, T; Yoshimoto, M. *Fuller. Sci. Technol,* 1996, 4, 599-612.
[3] Ikegami, T; Nakanishi, F; Uchiyama, M; Ebihara, K. *Thin Solid Films,* 2004, 457, 7-11.
[4] Ashfold, MNR; Claeyssens, F; Fuge, GM; Henley, SJ. *Chem. Soc. Rev.,* 2004, 33, 23-31.
[5] Yueh, FY; Singh, JP; Zhang, H. In *Encyclopedia of Analytical Chemistry*; Meyers RA. Ed. Laser-induced Breakdown Spectroscopy; Elemental Analysis; John Wiley & Sons Ltd, Chichester, 2000, pp 2066-2087.
[6] Dwivedi, RK; Thareja, RK. *Phys. Rev. B,* 1995, 51, 7160-7167.
[7] Demyanenco, AV; Letokhov, VS; Puretskii, AA; Ryabov, EA. *Quantum Electronics* 1998, 28, 33-37.
[8] Vivien, C; Hermann, J; Perrone, A; Luches, A. *J. Phys. D: Appl. Phys.* 1998, 31, 1263-1272.
[9] Wee, S; Park, S. M. *Opt. Comm.,* 1999, 165, 199-205.
[10] Yamagata, Y; Sharma, A; Narayan, J; Mayo, RM; Newman, JW. *J. Appl. Phys.,* 2000, 88, 6861-6867.
[11] Harilal, SS. *Appl. Surf. Science,* 2001, 172, 103-109.
[12] Acquaviva, S; Giorgi, ML. *J. Phys. B: At. Mol. Opt. Phys.,* 2006, 35, 795-806.
[13] Saito, K; Sakka, T; Ogata, H. *J. Appl. Phys.,* 2003, 94, 5530-5536.
[14] Zelinger, Z; Novotny, M; Bulir, J; Lancok, J; Kubat, P; Jelinek, M. *Contrib. Plasma Phys.,* 2003, 43, 426-432.

[15] Saidane, K; Razafinimanana, M; Lange, H; Huczko, A; Baltas, M; Gleizes, A; Meunier, JL. *J. Phys. D: Appl. Phys.*, 2004, 37, 232-239.
[16] Park, HS; Nam, SH; Park, SM. *J. Appl. Phys.*, 2005, 97, 113103-5.
[17] Fuge, GM; Ashfold, MNR; Henley, SJ., *J. Appl. Phys.*, 2006, 99, 14039-12.
[18] Camacho, JJ, Poyato, JML, Díaz, L; Santos, M. *J. Phys. B: At. Mol. Opt. Phys.*, 2008, 41,105201-13.
[19] Camacho, JJ; Santos, M; Diaz, L; Poyato, JML. *Appl. Phys. A.*, 2009, 94, 373-380.
[20] Camacho, JJ; Diaz, L; Santos, M; Juan, LJ; Poyato, JML. *J. Appl. Phys.*, 2009, 106, 33306-11.
[21] Cremers, DA; Radziemski, LJ. *Handbook of Laser-Induced Breakdown Spectroscopy*; Wiley: Chichester, England, 2006.
[22] Miziolek, AW; Palleschi, V; Schechter, I. (Eds.), *Laser-Induced Breakdown Spectroscopy;* Cambrige. 2006.
[23] Singh, JP; Thakur, SN. (Eds.) *Laser-Induced Breakdown Spectroscopy*; Elsevier: Oxford UK, 2007, Vol. 1, pp 1-427.
[24] Yong-Ill, L., *Laser Induced Breakdown Spectrometry*; Nova Science Publishers: New York 2000.
[25] Chan, F. *Introduction to plasma physics and controlled fusion*; Plenum Press: New York. 1984.
[26] Griem, HR. *Principles of plasma spectroscopy*; University Press: Cambridge. 1997.
[27] Griem, HR. *Phys. Rev.*, 1962, 128, 515-523.
[28] MacDonald, AD. *Microwave Breakdown in Gases* Wiley; New York, 1966.
[29] Raizer, YP. *Gas Discharge Physics;* Springer: Berlin, Heidelberg. 1991.
[30] Kopiczynski, TL; Bogdan, M; Kalin, AW; Schotwau, HJ; Kneubuhl, FH. *Appl. Phys. B: Photophys. Laser Chem.*, 1992, 54, 526-530.
[31] Radziemski, LJ; Cremers, DA; *Laser-induced plasma and applications*; New York: Dekker. 1989.
[32] Gurevich, A; Pitaevskii, L. *Sov. Phys. JETP,* 1962, 19, 870-871.
[33] Morgan, CG. *Rep. Prog. Phys.*, 1975, 38, 621-685.
[34] Huddlestone, RH; Leonard, SL. *Plasma diagnostic techniques*; Academic Press: New York. 1965.
[35] Hutchinson, IH. *Principles of plasma diagnostic*; University Press: Cambridge. 2002.
[36] NIST Atomic Spectra Database online at http://physics

[37] Breene, RG. *The Shift and Shape of Spectral Lines*; Pergamon: London. 1961.
[38] Bengtson, RD; Tannich, JD; Kepple, P. *Phys. Rev. A*, 1970, 1, 532-533.
[39] Griem, HR. *Spectral line broadening by plasmas*; Academic Press: New York.1974.
[40] Herzberg, G. *Spectra of diatomic molecules*; Van Nostrand: New York. 1950.
[41] Steinfeld, JI. *An introduction to modern molecular spectroscopy*; MIT Press: London. 1986
[42] Bernath, PF. *Spectra of atoms and molecules*; Oxford University Press: New York. 1995.
[43] Camacho, JJ; Pardo, A; Martin, E; Poyato, JML. *J. Phys. B: At. Mol. Opt. Phys.* 2006, 39, 2665-2679.
[44] Kovacs, I. *Rotational Structure in the Spectra of Diatomic Molecules*. Hilger: London. 1969.
[45] Cabalin, LM; Laserna, JJ; *Spectrochim. Acta Part B*, 1998, 53, 723-730.
[46] Demtröder, W. *Laser Spectroscopy*. Vol. 2. Experimental Techniques. Springer. Berlin 2008.
[47] Bogaerts, A; Chen, Z. *Spectrochim. Acta Part B,* 2005, 60,1280-1307.
[48] Zeldovich, YB; Raizer, YP. *Physic of Shock waves and high temperature hydrodynamics phenomena*. Academic, New York 1966.
[49] Drogoff, LB; Margotb, J; Chakera, M; Sabsabi, M; Barthelemy, O; Johnstona, TW; Lavillea, S; Vidala, F; Kaenela, VY. *Spectrochim. Acta Part B,* 2001, 56, 987-1002.
[50] Huber, KP; Herzberg, G. *Molecular spectra and Molecular structure. IV. Constants of diatomic molecules*; Van Nostrand Reinhold: New York. 1979.
[51] Martin, WC; Zalubas, R. *J. Phys. Chem. Ref. Data*,1983, 12, 323-380.
[52] Kim, DE; Yoo, KJ; Park, HK; Oh, KJ; Kim, DW. *Appl. Spectrosc.*, 1997, 51, 22-29.
[53] Lu, YF; Tao, ZB; Hong, MH. *Jpn. J. Appl. Phys.*, 1999, 38, 2958-2963.

INDEX

A

absorption, 8, 9, 12, 13, 22, 27, 33, 44, 51, 65
absorption coefficient, 22
acceleration, 13
access, 29
accuracy, 20
acetylene, 38, 43
acoustic, 33
activation, 30
aid, 29, 60
air, 2, 9, 35, 37, 38, 39, 43, 44, 45, 46, 48, 49, 51, 52, 53, 57, 62
ambient pressure, vii, 2, 71
angular momentum, 18, 25
anode, 28
arsenide, 27
assignment, vii, 18, 35, 38, 39, 43, 44, 71
assumptions, 44
atmosphere, 35, 45, 51, 53
atmospheric pressure, 9, 27
atoms, 4, 5, 7, 9, 10, 12, 14, 16, 19, 22, 23, 26, 33, 45, 47, 68, 69, 77
attachment, 9, 10, 11
averaging, 62, 63

B

back, 30, 33
base, 29
beams, 11, 34
behaviours, 65
bending, 25
binding, 45
binding energy, 45
Bohr, 6
Boltzmann constant, 6
Boltzmann distribution, 44
breakdown, vii, 2, 3, 8, 9, 10, 14, 20, 21, 33, 45, 51, 53, 59, 60, 62
bremsstrahlung, 9, 13, 33, 47
burning, 37, 38
butane, 37, 38

C

calibration, 30
carbon, 1, 2, 25, 29, 35, 37, 38, 39, 44, 45, 46, 47, 48, 49, 50, 51, 52, 53, 57, 59, 60, 62, 64, 65, 66, 67, 69
carbon atoms, 47
carbon dioxide, 25, 48, 50
cathode, 28, 30

cavitation, 33
cell, 29
charge coupled device, 28
charged particle, 4, 5, 13
chemical, iv, 1, 2, 4, 5, 17, 19, 54
chemical reactions, 4
classical, 8, 12
clusters, 45
CO_2, i, iii, iv, vii, 2, 25, 26, 29, 30, 33, 34, 35, 37, 38, 39, 43, 44, 45, 47, 49, 51, 54, 59, 60, 62, 65, 66, 67, 69, 71
collisions, 1, 9, 10, 16, 22, 46, 47, 62, 67, 71
color, iv
combined effect, 11
combustion, 43
complexity, 40
components, 4, 13, 43
composition, 1, 5, 17, 51
concentration, 10
conductive, 4
configuration, 7, 27, 29
confinement, 48
continuity, 9
control, 12, 31
cooling, 30, 59, 67, 68
copper, 4, 27
corona, 4
Coulomb, 5, 12, 13
coupling, 18
cracking, 33
cyclotron, 13

D

database, 41
decay, 5, 14, 55, 69
degenerate, 25, 44
Delta, 29
density, vii, 4, 5, 6, 8, 9, 10, 11, 14, 15, 16, 17, 20, 21, 22, 23, 29, 38, 42, 46, 49, 57, 59, 66, 68, 69, 71
deposition, 1

depth, 12, 13
detection, 17, 20, 25, 27, 28, 30, 43, 55
detection system, 17, 25, 27, 30, 55
diaphragm, 29, 34
diffraction, 21, 29, 36
diffusion, 10, 11
distribution, 5, 11, 14, 44
divergence, 21, 29, 34
Doppler, 14, 46
duration, 15, 20, 21, 29, 33, 67
dynamical properties, 2

E

earth, 4
electric arc, 38
electric charge, 4
electric field, 8, 9, 15, 16, 20, 35
electrical properties, 4
electromagnetic, 4, 8
electromagnetic fields, 4
electromagnetic wave, 8
electron, vii, 2, 4, 6, 7, 8, 9, 10, 11, 12, 13, 14, 16, 19, 22, 26, 46, 47, 57, 59, 66, 68, 69, 71
electron charge, 8
electron density, 4, 6, 8, 9, 10, 11, 46, 59, 67, 68, 69
electrons, 4, 5, 6, 7, 9, 10, 11, 12, 22, 28, 45, 47, 53, 66, 67, 69
emission source, vii, 71
emitters, 15, 71
energy, 1, 3, 4, 5, 6, 7, 8, 9, 10, 12, 13, 14, 15, 17, 18, 20, 21, 22, 23, 26, 29, 42, 43, 44, 45, 46, 47, 51, 53, 65, 67
energy density, 4, 22
energy transfer, 26
England, 76
environment, 12
equilibrium, 4, 5, 7, 17, 19, 47
estimating, 18
evaporation, 22

evolution, 1, 5, 7, 51, 59, 62, 67, 68, 69
excitation, 2, 3, 9, 17, 22, 26, 39, 40, 46, 54, 66, 67
experimental condition, 39

F

flame, 37, 38, 43
flight, vii, 62
fluctuations, 20, 21
fluorescence, 4, 25
fluorescent lamps, 4
focusing, 3, 20, 27, 55
force, 15
formation, 2, 5, 11, 21, 33, 47, 54, 67
formula, 8
Fox, 35, 36, 59
fragments, 2, 36, 51
free energy, 13
freedom, 9
fusion, 76
FWHM, 14, 15, 16, 20, 29

G

GaAs, 27
gallium, 27
gas, 1, 2, 3, 4, 5, 14, 15, 19, 20, 25, 26, 33, 46, 53, 54, 57, 65, 67, 71
gas phase, 54, 57, 67
gases, 19
gauge, 29
Gaussian, 15, 21
generation, 8, 9, 10
geometry, 33
germanium, 27
gold, 4, 27
graph, 46
graphite, vii, 2, 29, 33, 36, 39, 43, 45, 54, 55, 56, 59, 60, 61, 62, 63, 65, 67, 71
gratings, 30
growth, 8, 9, 10, 11, 65

H

halogen, 30
heat, 26
heat capacity, 27
heavy particle, 67
height, 4
high pressure, 48, 51
high temperature, 5, 33, 45, 77
history, 59, 60
HK, 77
HR, 76, 77
hydrocarbon, 43
hydrodynamics, 77
hydrogen, 6, 16, 23, 36, 40

I

IB, 22
image, 25, 27, 33
images, 27
incidence, 59, 64
indication, 62
industry, 4
inelastic, 9, 47
infrared, vii, 1, 25, 27
insertion, 30
insight, 4, 39, 62
instabilities, 5
instruments, 28
integration, 12, 13, 30, 47
intensity values, 17
interaction, 1, 15, 16, 20, 47
interactions, 1, 12, 48
interval, 30, 62, 65
intrinsic, 14
inversion, 26, 27
ionic, 2, 3, 6, 7, 13, 16, 17, 22, 36, 38, 44, 47, 51, 54, 55, 59, 60, 62, 63, 65, 66, 67, 69
ionization, vii, 2, 6, 7, 8, 9, 10, 11, 12, 13, 19, 22, 23, 45, 46, 47, 71

ionization energy, 9, 10
ionization potentials, 12, 46
ionosphere, 4
ions, 4, 5, 6, 7, 9, 10, 12, 14, 19, 45, 47, 66, 67, 68, 69
$_{IP}$, 8, 9, 19, 45, 46
IR, vii, 1, 2, 9, 27, 54, 59, 71
irradiation, 29, 60

J

JI, 77

K

kinetic energy, 7, 9, 22, 67

L

LA, 35, 39, 44, 49, 52
laser, vii, 1, 2, 3, 7, 8, 9, 10, 11, 14, 20, 21, 22, 25, 27, 28, 29, 30, 33, 34, 35, 37, 38, 39, 42, 43, 44, 45, 46, 47, 48, 49, 50, 51, 54, 59, 60, 62, 63, 64, 65, 66, 67, 69, 71
laser ablation, vii, 1, 2, 28, 54, 55, 59, 62, 63, 67
Laser Induced Breakdown Spectroscopy, 3
Laser Induced Breakdown Spectroscopy (LIBS), 3
laser radiation, 1
lasers, 4, 22, 25, 27, 65
law, 5
lead, 21, 33
lens, 21, 29, 55
LIBS, vii, 2, 3, 12, 16, 20, 22, 25, 27, 28, 29, 34, 59, 71
lifetime, 14, 15
light, 3, 9, 20, 21, 27, 29, 33
light beam, 27
limitation, 67
linear, 1, 15, 16, 25, 40, 42
LM, 77

local conditions, 14
London, 19, 77
losses, 11, 22, 29
low power, 7
low temperatures, 16

M

magnetic, 13
magnetic field, 13
magnitude, 6
maintenance, 27
man-made, 4
mass, 1, 6, 15
materials, 1, 27
matrix, 30
matter, 3, 4
measurement, 21, 67
measurements, 20, 29, 48, 59, 62, 66
melt, 45
microwave, 8
military, 25
mirror, 27
mission, 55, 67, 69
missions, 28, 30
MIT, 77
ML, 75
molecules, 4, 5, 7, 9, 10, 12, 14, 26, 45, 68, 71, 77
molybdenum, 27
momentum, 18, 25
monochromator, 28
motion, 14
MPI, 8, 9, 10, 11, 47
MS, 75
multiplication, 28
MV, 34

N

NaCl, 27, 29
natural, 4, 9, 14

n_D, 16
neutral, 4, 7, 9, 10, 12, 13, 16, 19, 22, 45
New York, 76, 77
N_i, 6, 19, 22, 46, 47
NIR, vii, 35, 71
NIST, 13, 18, 38, 39, 41, 76
nitrogen, 62
noise, 28, 30, 33, 67
non-thermal, 47
normal, 25, 26, 29

O

observations, 33
OH, vii, 2, 35, 36, 39, 45, 48, 50, 51, 54, 55, 60, 69, 71
online, 76
opportunities, 4
optical, 9, 18, 21, 25, 27, 28, 29, 30, 62, 67
optical pulses, 25
optics, 29
orbit, 9
oscillation, 8
oscillations, 26
oscillator, 12
overlap, 39
oxygen, 37, 38, 43, 62

P

parabolic, 27
parallel, 27, 55
parameter, 6, 16, 66
parity, 44
particle collisions, 47
particles, 1, 2, 4, 5, 12, 13, 15, 16, 54
partition, 17
permission, iv
permit, 4
PF, 77
photoionization, 13, 22

photon, 7, 9, 12, 13, 14, 20, 22, 23, 29, 30, 34, 35, 60
photonics, 28
photons, 4, 8, 10, 47
physical properties, 4, 20
physicists, 5
physics, 4, 76
Planck constant, 20
plane waves, 21
plasma physics, 76
play, 22, 26
PMI, 47
poor, 21, 67
population, 5, 6, 7, 12, 19, 27
population density, 12
power, vii, 2, 3, 7, 8, 12, 20, 21, 25, 27, 29, 33, 35, 38, 42, 46, 47, 48, 49, 54, 59, 71
preparation, iv
pressure, vii, 2, 9, 14, 15, 27, 29, 35, 37, 38, 39, 42, 43, 44, 45, 46, 47, 48, 49, 51, 53, 71
probability, 8, 12, 13, 14, 17, 47
production, 71
program, 18
propagation, vii, 1, 30, 33, 54, 57, 71
propane, 38
pulse, 1, 3, 7, 20, 21, 22, 27, 28, 29, 30, 33, 35, 47, 59, 60, 62, 64, 65, 66, 67
pulsed laser, vii, 1, 2, 4, 25, 28, 54, 55, 59, 63, 71
pulsed laser deposition, 1
pulses, 25, 65
pumping, 65
purity, 29

Q

quanta, 25
quantification, 14
quantum, 6, 7, 39, 44
quartz, 29

R

RA, 75
radial distribution, 11
radiation, 1, 3, 12, 13, 14, 15, 27, 28, 33
Radiation, 12, 13
radicals, 4
radius, 6, 11, 21, 29
Raman, 3
Raman spectroscopy, 3
range, 3, 5, 12, 16, 26, 27, 28, 29, 36, 44, 46, 51, 59, 62, 67
reactions, 2, 4, 19, 54, 57
recombination, 7, 10, 11, 13, 19, 57, 66, 67
recombination processes, 67
recommendations, iv
reflectivity, 27, 29
regression, 40
regular, 39
relationship, 12
relaxation, 2, 36, 40, 51
relaxation process, 40
resolution, 14, 30, 35, 36, 37, 38, 39, 44, 46, 49, 52, 59, 60, 62, 63, 64, 65, 66
resonator, 26
response, 4, 17, 30
RF, 75
rights, iv
rings, 21
room temperature, 4
routes, 71
rules, 44

S

sample, 1, 3, 20, 21, 48
sampling, 1, 4
satellite, 44
selecting, 17
semiconductor, 28
sensitivity, 3, 28, 55
separation, 10
series, 17, 21, 36
SH, 76
shape, 14, 15, 18, 21, 29, 34, 59, 60
shock, 33, 68
shock waves, 33
sign, 16
signals, 28
signal-to-noise ratio, 28, 67
silicon, 28
simulation, 18
simulations, 19
skin, 11
Spain, 73
species, iv, vii, 1, 2, 3, 4, 5, 7, 9, 10, 11, 12, 16, 17, 18, 36, 46, 47, 53, 54, 55, 57, 59, 60, 62, 65, 67, 69, 71
spectral component, 27
spectroscopic methods, 25, 71
spectroscopic techniques, 4, 11
spectroscopy, iv, vii, 1, 2, 3, 12, 25, 27, 76, 77
spectrum, 2, 18, 35, 36, 39, 43, 49, 59, 60, 62, 64, 69
speed, 4, 20
speed of light, 20
stages, 62
Stark effect, 12, 15, 16, 69
state, 3, 4, 5, 6, 7, 12, 13, 14, 19, 22, 23, 25, 43, 47
states, 4, 9, 12, 17, 18, 22, 26, 44, 45, 69
strength, 12, 13, 15
stretching, 26
strikes, 27
structure, 6, 12, 18, 36, 39, 43, 44, 63, 77
superposition, 63
suppression, 65
surface component, 53
surgical, 25
symmetry, 25
synchronization, 30
synthesis, 4

Index

T

target, vii, 1, 2, 22, 29, 30, 33, 45, 47, 48, 53, 54, 55, 56, 57, 59, 62, 65, 66, 67, 71
targets, 59, 67
techniques, 3, 11, 76
technologies, 1
temperature, 2, 4, 5, 11, 13, 14, 15, 16, 17, 18, 19, 22, 29, 40, 41, 43, 44, 45, 46, 53, 59, 66, 68, 77
temperature dependence, 22
temporal, 7, 11, 29, 59, 60, 62, 65
textbooks, 3
thermal energy, 67
thermal expansion, 33
thermodynamic, 4, 5, 7, 14
thermodynamic equilibrium, 4, 5, 7, 14
thermonuclear, 4
threshold, 10, 13, 21, 33, 65
time use, 47
transfer, 26
transition, 6, 12, 13, 14, 17, 18, 19, 39, 40, 41, 43
transitions, 12, 13, 18, 26, 40
transmission, 33
transparent, 27
travel, 30
two-dimensional, 28

U

UK, 76
ultraviolet, vii
uncertainty, 14, 17, 46
UV, vii, 9, 35, 62, 65, 71

V

vacuum, vii, 1, 29, 35, 42, 45, 48, 71
validity, 7
values, 5, 13, 17, 39, 46, 66
Van der Waals, 15
vapor, 45
variables, 20
variation, 43, 67
velocity, 6, 11, 14, 22, 62
vibration, 25
vibrational modes, 26
visible, vii, 9, 27, 43, 65

W

water, 29
wavelengths, vii, 9, 47, 71
weapons, 25
welding, 25
wind, 4
windows, 29

Z

zinc, 27